CLOSED CIRCUITS

CLOSED CIRCUITS

The Sellout of Canadian Television

HERSCHEL HARDIN

Douglas & McIntyre
Vancouver/Toronto

Douglas & McIntyre Ltd.
1615 Venables Street
Vancouver, British Columbia V5L 2H1

Canadian Cataloguing in Publication Data

Hardin, Herschel, 1936–
    Closed circuits

ISBN 0-88894-446-2

1. Television broadcasting – Canada – History.
I. Title.
HE9700.9.C2H37   1984   384.55'4'0971
                                            C84-091390-7

Jacket illustration by Ron Lightburn
Design by Barbara Hodgson
Typeset by Ronalds Printing
Printed and bound in Canada by Imprimerie Gagné Ltée.

Contents

Part I

Great Expectations: The Creation of the CRTC

1

Noble Documents
THE FOWLER COMMITTEE
REPORT OF 1965

*The only thing that really matters in broadcasting is program content;
all the rest is housekeeping.*

—The Fowler Committee Report

ON SEPTEMBER 1, 1965, a special advisory committee on broadcasting, known as the Fowler Committee after its chairman, Robert Fowler, submitted its report to Secretary of State Maurice Lamontagne. Chairman Fowler was an Ontario lawyer stationed in Montreal as president of the Pulp and Paper Institute of Canada. In effect he was undertaking a sequel to the report of the Royal Commission on Broadcasting of 1955–57, of which he had also been chairman. Fowler, in his off hours, sometimes helped out as à Liberal party adviser, with direct access to Prime Minister Lester Pearson. "Bob Fowler was extremely useful in proposing new ideas, whether for policy or organization," Pearson was to note later, writing of efforts to rebuild the Liberal party.

One of the committee's two other members was a Montreal lawyer, Marc Lalonde, a federalist friend of law professor and theoretician Pierre Trudeau, and also a friend of Lamontagne. Lalonde and Trudeau, together with five other Quebeckers, had not too long before issued a Manifesto for the Nation supporting the federalist option. Ernest Steele, the committee's third member, was undersecretary of state and a former secretary of the Treasury Board, and also a friend of Lamontagne.

The Fowler Report was a noble document and a clarion call:

When we declare that broadcasting should be a major instrument for the development of a distinctive Canadian culture, we use that most abused word "culture" in its broadest and original meaning. It is the reflection of life itself, in all its variety — its beauty and ugliness; its significant artistic achievements and its unimportant daily occurrences; its big people and its little people; its important and often inscrutable messages, and its light insignificant interests; its great opinions and its amusing anecdotes; tragedy and comedy, laughter and tears, criticism, irony, satire, and sheer fun and amusement — all are essential.

To reflect a nation's culture — and to help create it — a broadcasting system must not minister solely to the comfort of the people. It must not always play safe. Its guiding rule cannot be to give the people what they want, for at best this can be only what the broadcasters think the people want; they may not know, and the people themselves may not know. One of the essential tasks of a broadcasting system is to stir up the minds and emotions of the people, and occasionally to make large numbers of them acutely uncomfortable. Of course broadcasters should not all the time play the role of an Old Testament prophet; most of the time they are engaged in providing easily digestible and agreeable fare. But, in a vital broadcasting system, there must be room for the thinker, the disturber, and the creator of new forms and ideas. He must be free to experiment — to fail as well as to succeed in the expression of new ideas.

Each chapter of the report was headed with a quotation, usually from a literary source. Lewis Carroll, Boris Pasternak, Jane Austen, Thomas Mann, Molière, Tennyson, Samuel Clemens and Beaumarchais, among others, made their appearance. In the English edition of the report, the French citations appeared in French, a nice touch.

The report was replete with facts and analysis. It described the contribution of private television to Canadian programming and to the support of Canadian artists as "contemptuous." It detailed the Americanization of Canada's television, particularly in prime-time hours. In the Toronto area, for example, two-thirds of evening programs of English-language commercial stations were of American origin, and as many as 75 per cent of viewers of those stations, in the evening, were watching the U.S. material. The Canadian operators could pick up the American programs for a song — 5 to 8 per cent of the cost of production. "Let to operate freely, economic factors would quickly tend to make Canadian private television stations mere extensions of the American networks."

Damned if the licensing of additional Canadian but commercial stations (the CTV stations) hadn't in general increased the viewing of American programs!

The advent of these stations, moreover, instead of widening the scope of programs available to Canadian viewers, had "merely increased the broadcasting of popular entertainment, mainly of American origin." The imbalance of programming on these stations was striking. "The concept of program balance requires broadcasters to expose the public to a wide range of experience," the committee explained. During the peak hours, 7 P.M. to 10 P.M., on the other hand, 92.9 per cent of CTV programming was entertainment. Statistical tables and impressive coloured bar graphs were provided. Of the Canadian programming that was produced, there had been "too many quiz shows and similar types of programs" to fill the loose Canadian content requirements. Canadian television as a whole might be "mature physically," but it was not "mature mentally."

In fact, the program performance of the commercial stations — in particular the additional second stations — bore "very little relationship to the promises made to the Board of Broadcast Governors [the regulatory agency at the time] when the licences were recommended. Undertakings given to obtain the grant of a public asset have largely been ignored." A promise made by a broadcaster to obtain a licence to operate a radio or television station "should be an enforceable undertaking," the report complained, "and not a theoretical exercise in imagination or a competitive bid in an auction of unrealistic enthusiasm."

Well, Canadian television was in quite a multitudinous mess. The committee now proposed a new regulatory agency with instructions to get tough — to lay down the law. Promises of applicants should be made contractual conditions of licences, enforceable by prosecution or suspension of the licences. Television channels were valuable public assets. "No one has the automatic right to the 'renewal' of a licence." Although the report did not say it in so many words, hadn't the commercial operators betrayed Robert Fowler and his royal commission of 1957 which had recommended their licensing? The committee wasn't going to take it any more.

The report of the Fowler Committee eventually led to the Broadcasting Act of 1967–68. Section 3 of the new act was eloquent in catching the spirit of the committee's hopes and objectives. Section 3(b) stated that "the Canadian broadcasting system should be effectively owned and controlled by Canadians so as to safeguard, enrich and strengthen the cultural, political, social and economic fabric of Canada." Section 3(d) called for the use by each broadcaster of "predominantly Canadian creative and other resources." Section 3(a) stated that "broadcasting undertakings in Canada make use of radio frequencies that are public property." These stirring clauses, rich in historical and public-interest significance, would subsequently be quoted many times. They were particularly good for subtitles on reports, or for speeches by ministers of communications, after the introductory jokes.

The whole of Section 3 was entitled "Broadcasting Policy for Canada." Most legislation does not even bother with a policy preamble. Section 4 of the act, in turn, created the Canadian Radio-Television Commission (CRTC) to replace the discredited Board of Broadcast Governors (BBG). The first chairman of the new agency was Pierre Juneau, a Montrealer whom Judy LaMarsh, then secretary of state, had persuaded to leave the National Film Board for the BBG, with the chairmanship of the envisaged new agency in mind. Juneau, as BBG vice-chairman, had already launched into action, tripling the staff of the BBG, putting together an advisory committee on programming and hiring a firm of chartered accountants to analyze the financial results of the commercial stations. "This was my first step in my plan to bring qualified, non-partisan people to broadcasting," LaMarsh recounted, and even at this prior stage, "Juneau proved my faith in him many times over."

For vice-chairman to Juneau, LaMarsh appointed Harry Boyle, a program supervisor at the CBC in Toronto. The other three full-time members were Pat Pearce, television columnist for the Montreal *Star,* Real Therrien, a Quebec City broadcasting and communications consultant, and Hal Dornan, originally from Vancouver, a former press secretary to the prime minister. Among the part-time commissioners was Northrop Frye, the internationally known literature scholar from the University of Toronto. Jacques Hébert, a Montreal publisher and an old friend of the soon-to-be prime minister, Pierre Trudeau, was also later to join the commission as a part-time member.

Juneau, Boyle, Pearce, Frye, Hébert — culturally-minded and cultured people, alert, aware and sensitive, quite obviously. The new agency could be said to have a definite touch of class.

2

The Canadian-Content Regulations of 1970

IN THE MARCH 1970 issue of *Saturday Night*, the magazine of the liberal well-educated in southern Ontario and anglophone Montreal, editor Robert Fulford published, under his regular personal by-line, an article entitled "Juneau's Revolution: Making TV History." This was two years after the creation of the CRTC but, historically speaking, the time gap was only a moment. Juneau had just announced proposed new Canadian-content rules, which prompted the Fulford article.

If the creation of the CRTC in 1968 was an Immaculate Liberal (Very High Frequency) Conception, then "Juneau's revolution" was its liberal intellectual baptism, in the authorized *Saturday Night* version. For years, through royal commissions and parliamentary committees, the article recounted, "everyone had been saying that the Canadian broadcasting system should be Canadian." And yet, through all those years, the Canadian broadcasting system grew more and more American. The advent of television, so much more expensive than radio production, had "drowned" the CBC with Hollywood programs. The private television stations, for their part, "made little more than a few gestures toward Canadian programming, and the private radio stations hardly did that."

"And then, on February 12, 1970, the Canadian Radio-Television Commission announced that this period was at an end." The commission had proposed that Canadian television fill 60 per cent of its prime-time hours with Canadian programs and that 30 per cent of all music on radio stations would have to be Canadian. It seemed, in the description, like the second day of Genesis. This announcement made Juneau "obviously ... the most important single figure in Canadian mass culture," Fulford went on. The implications of the proposal were "enormous." Juneau was actually planning to "order various Canadian institutions to do what everyone had always said they should do." Pokes were also taken at the commission's predecessor, the Board of Broadcast Governors, "amateurs of broadcasting who knew little about programming and believed that when they announced percentage goals their most serious work was done." The commission, on the other hand, was "radically different. . . . When the CRTC meets a broadcaster it meets him on terms of professional equality. It offers not empty rhetoric but opinions based on detailed facts." The commission had, in particular, studied "the economics" and knew not only what "*should* be done" but what "actually *could* be done and *would* be done."

There were other signs that the commissioners took "their work much more seriously than most regulatory agencies." They had cancelled the licence of a small radio station, for cause. By comparison, the Federal Communications Commission in the United States, in four decades, had "never seen fit to take such an obvious step."

When the commission turned to cable television and announced that cable "was not to have a free ride on the back of broadcasting," they really "began to make broadcasting history." Cable operators "were expected to create their own programmes [a reference to the community channel proposal] and put back into their communities some of the money they were taking out."

> Immediately the commissioners changed the status of cable TV from yet another licence to print money (in Lord Thomson's famous phrase) to the status of a public utility with public responsibilities. About the same time, the CRTC began to make it clear that it believed broadcasting stations should be owned in their communities and should reflect the needs of their communities.
>
> . . . In a field demoralized on the one hand by a gone-to-seed public corporation, and corrupted on the other by the deadest sort of commercialism, they were calling for hope, optimism and national ambition on a grand scale. In the middle of a dull, anxious winter, the February announcements of the Canadian Radio-Television Commission were a remarkable sign of life. The idea of Canadian broadcasting, as it turned out, wasn't dead yet.

The next few months of 1970 were heroic for the CRTC. The commission's Canadian-content proposals were described in the press as "bold philosophy." Reports heralded the "historic hearing" to consider the proposals — "the commission's single most important confrontation to date" — that was scheduled for April 1970. "Our TV future at stake as broadcasters prepare to fight new rulings," ran a headline in the Toronto *Star*. For the making of illusion, the events of the spring of 1970 would be hard to beat anywhere.

To begin with, a real historical fight in broadcasting politics was still taking place. The Canadian broadcasting tradition of self-expression and of public service that Juneau was trying to defend was not just a bright idea that Juneau, Boyle and the others had concocted in a moment of enthusiasm. The tradition went back to the late 1920s, when Canadian National Railways established the country's first regular network-radio broadcasts, as a public-broadcasting service. The Canadian Radio-Broadcasting Commission (CRBC) of 1932 and its successor, the Canadian Broadcasting Corporation (1936), followed. Prime Minister R. B. Bennett, in introducing the 1932 legislation, declared, in support of public ownership of broadcasting, "This country must be assured of complete Canadian control of broadcasting from Canadian sources, free from foreign interference or influence. . . . I cannot think that any government would be warranted in leaving the air to private exploitation and not reserving it for development for the use of the people." Bennett's speech was just one of many thoughtful contributions in favour of a public system, with a public-service mandate, individualized programming and democratic ownership.

When Bennett and others talked about reserving the airwaves for the development and use of the people, rather than for selling commercials, they had more than these classic objectives of public broadcasting in mind. Short of the public-broadcasting initiatives of the 1930s, Canadian radio would have fallen into the hands of the U.S. networks. The takeover had already begun. Nationalism — keeping the country together, maintaining Canada's own traditions and self-expression next to the United States — was always at the centre of the debate.

The 1929 Royal Commission on Broadcasting (the Aird Commission), which had led to the 1932 legislation, had stated equally clearly that public ownership and financing were both necessary and desirable. The years 1929–36 also saw the rise of the legendary Canadian Radio League (headed by Graham Spry and Alan Plaunt), a citizens' organization that led the debate for a public-broadcasting system in Canada. Public discussion was widespread. Radio was a great issue of the day, and the questions of who should control it and how it should be used were widely explored.

The 1932 decision for public broadcasting actually produced the desired Canadian effect, moreso than anyone had ever imagined. The dramatic advances of the early CBC also demonstrated the general practical value of public broadcasting; Canadians could have their own distinctive broadcasting suited to their own circumstances, despite what the leading commercial-radio interests had said in arguing for a sellout to the American networks. Most of all, this public-broadcasting tradition was a vital expression of the country's different, un-American identity.

The full story of the rise of public broadcasting in Canada, which could as easily be called the rise of Canadian broadcasting in Canada, is told by Frank Peers in *The Politics of Canadian Broadcasting 1920–1951* and by Austin Weir in *The Struggle for National Broadcasting in Canada*. There is also an analysis of the tradition, by myself, in *A Nation Unaware*.

At the CRTC's 1970 Canadian-content hearings, the situation facing the commission in television was almost identical to that facing radio in the 1920s. The latter had led to the original public broadcasting response. Now Canadians were watching mostly American programs by virtue of their greater availability and resources. The same kind of lopsided percentage figures had been trotted out in the early radio days.

Similarly, raise the overall broadcasting question, as the hearings did, and one was driven to talk about the country itself. Juneau's "It is a matter of deciding whether we want a Canadian broadcasting system or we do not" would inevitably become "Either we have a country or we don't." Vice-chairman Harry Boyle worried out loud about whether the extra U.S. program distribution on cable was destroying the linkage holding Canada together. "Already it takes great bites out of that blood flow from east to west. If you allow it to go on it will gobble up the whole system. We've got to find a means of reinstituting the same principle of keeping an east-west flow."

In addition, in 1970 the record of the first decade of Canadian commercial television stations was in, and it was damning. Expo and growing economic and cultural nationalism had created renewed criticism of wholesale Americanization and renewed feelings of Canadian self-affirmation here and there across the country.

On the other side were the commercial broadcasters. They had maintained a foothold in radio in the 1930s, and then expanded, and then moved into television, virtually splitting the broadcasting system and undermining the Canadian objectives of broadcasting legislation. Their convenient ideology was the American ideology that freedom means the freedom of commercial exploitation, and their freedom to be as American, conformist, derivative and pandering as humanly profitable, while protesting nevertheless their Canadian citizenship. The more that Juneau and Boyle talked turkey to encourage resolve and a sense of challenge, the greater became the commercial operators' indignation and ideological cant. They rallied their forces against the upcoming Canadian-content hearing and the new regulations.

The *Broadcaster*, the commercial industry's trade magazine, starting with its March 1970 issue, called for a religious war. A special two-page editorial was captioned: "No Virginia, a Canadian is not free to decide what he will watch on television or listen to on radio." The way that the CRTC was implementing new Canadian-content regulations, without consulting the public or the people's representatives in the House of Commons, was "a shaft pointed right at the heart of the democratic system which is our national life blood," the editorial thundered, not forgetting to call the CRTC "a group of fanatical idealists" while it was at it.

The cover of the April issue showed a television set and a radio padlocked with a ball and chain, marked "Property of CRTC."

> During the war, Hitler recognized German radio as a deadly weapon he could use, first to conquer his own country and bend its people to his will, and later to subjugate a whole continent.
>
> With press and radio muzzled, the stifling of other private enterprise was only a matter of time.
>
> As everyone knows, today, in the Communist countries, government control of information and all forms of cultural endeavor is used to dominate people.
>
> What [the CRTC] is really doing   perhaps unknowingly — is laying the foundations on which other less scrupulous authorities may readily build a fascistic machine, comparable to the one Hitler used to enslave most of Europe.

Space for this cover story was found below another article on the regular Finance page. "CRTC activity continues to disturb investor confidence," said the headline.

Just in case readers did not get the message, the editorial in the same issue demanded that "The CRTC rampage must be stopped," a rampage "aimed at the total subjugation of the Canadian broadcast media." The magazine reprinted an article from a 1946 issue, originally entitled "Can't Happen Here, Eh?" The article was a variation on George Orwell's *1984,* featuring the Department of Knowledge Control, 167 "People's Corporations" (including the Canadian Religious Corporation), and the elimination in the House of Commons of what used to be called the "Opposition Benches." This article was originally aimed at the CBC which, in 1946, was responsible for broadcasting regulation.

The outgoing president of the Canadian Association of Broadcasters (CAB), the commercial broadcasters' trade lobby, accused the CRTC of trying to build a wall against the United States and, for good measure, accused the CBC of spending 90 per cent of its budget buying U.S. programs. One newspaper columnist suggested that Sibelius, Mozart, Bach, Beethoven, Shostakovich, Debussy, Handel and Brahms would never be heard in Canada unless they could get their stuff recorded by Don Messer and his Islanders. Another columnist accused the CRTC of inward-looking isolationism and of clamping down the curtain.

CTV president Murray Chercover, worried apparently almost to the point of anguish, attacked the proposals for aiming at tonnage rather than quality. Viewers would switch to U.S. channels and undermine CTV's feasibility. The proposed new regulations would run CTV into the ground, turning a $4 million profit into a $9 million loss.

The Canadian Association of Broadcasters went even further, claiming that there just was not enough Canadian talent. Canadian music could not compete. There would be so many dull Canadian programs on television, viewers would flee. The proposals would be a disaster.

Prior to the hearing, vice-chairman Harry Boyle had replied to the doomsayers, "If you create opportunity, talent appears. But if it's not there, you'll never know by doing nothing." The complaints by the industry about the lack of money and talent were "the same old story" trotted out every time the industry was exhorted to pull up its socks.

"If we're as bad as this, what the hell are we doing here?" Boyle exclaimed. "If we don't tackle this thing then we might as well pack up now . . . Let [Parliament] change the act and we'll become a rebroadcasting system." These were challenging and inspiring words.

At the hearing itself, the CAB undertook an unprecedented seven-hour filibuster, lashing out every which way, including a bitter personal attack on a newspaper writer (Patrick Scott of the Toronto

Star) whose opinion of the CAB and its arguments was not what the freedom-loving CAB was prepared to tolerate. A CAB spokesman would later claim that the hearing was dominated not by the commission or by broadcasters, but by certain newspaper columnists.

The CAB's position was too much for two of the association's most powerful members, John Bassett (CFTO Toronto) and Stuart Griffiths (CJOH Ottawa), who resigned with a formal statement dissociating themselves from the CAB submission's "completely negative attitude."

Undaunted, if not undisturbed, the CRTC held its ground.

As for the CAB submission and filibuster, Juneau, according to Patrick Scott, cut it to shreds with surgical precision, "mainly by way of a virtuoso performance that laid bare the association's pathetic ignorance of all the Canadian performing talent it had said did not exist." One of the leading spokesmen for the broadcasters even admitted that he had never heard of Maureen Forrester. By the time Juneau had finished, "the befuddled broadcasters had backed down on every point in their brief. If it hadn't been so funny, as well as so richly deserved, it would have reduced a grown man to tears." The CAB's "frantic filibuster against the new content quotas . . . turned into a hollow pumpkin" as Juneau "dismantled its own high content quota of myths and half-truths."

CTV and the put-upon Mr. Chercover didn't fare much better. When Chercover emphasized that money was the key and that CTV did not have enough of it to match in quality what the commission wanted in quantity, Juneau zeroed in. "Let's just put that one to bed right now, Mr. Chercover," he responded. "The funds of the network are provided by its members, and we on the commission know what funds you have available."

Harry Boyle, putting down his omnipresent pipe, cut to the bone. "We have been getting," he said, "an awful lot of negative flak throughout this hearing about what can't be done in the way of Canadian programming under our regulations. Now tell me: If there were no regulations at all, would there be any Canadian programming?"

The Association of Canadian Television and Radio Artists (ACTRA) sent a heavyweight delegation in favour of the quotas, among them Fred Davis, Bruno Gerussi, Pierre Berton, Warren Davis, Betty Kennedy, Lorraine Thomson, Adrienne Clarkson, Farley Mowat, Hugh Garner, Frances Hyland and Don Harron. Pierre Berton called the hearing the most important meeting for the future of Canada that he had ever attended.

Juneau at one point commented that "we all seem to be discussing the broadcasting system as though it were a labyrinth whose corridors we can only narrow or widen. But shouldn't we be talking about getting out of the labyrinth altogether? Shouldn't we be looking for a whole new system?" The automatic scheduling of U.S. programs in prime time which guarantees them top ratings, and the reverse downgrading of Canadian possibilities, had become "a habit, a system that we can still break out of."

The commission's resolve quite noticeably stiffened as the hearing progressed. From time to time the commissioners betrayed disgust, contempt, rising temperatures and other negative symptoms over the industry's tactics and mentality. The Canadian Association of Broadcasters fell — some would say jumped — into disrepute. The filibustering tactic — the politically bizarre assumption that they could talk the commission into submission — instead infuriated the commission.

The week-long hearing provided copy for journalists day after day. Patrick Scott of the Toronto *Star,* perhaps the best journalist on the spot, began his wind-up column with the observation that "one could write a book about the latest landmark hearing . . . and maybe one day, when a few of the principals are dead, somebody will." This deference to the living, from a columnist who usually showed no deference to anything or anybody, provides a sense of the occasion.

The commission had made it plain that while they might amend the proposals, given solid reasons to do so, they intended to change the old rules, and in no mere token fashion. Now, with the hearing over, it was clearer than ever that change was on the way.

Two weeks after the hearing, Juneau appeared with other commissioners before the parliamentary committee on broadcasting. The encounter was billed as a briefing session on the Canadian-content proposals. It turned into a stormy two-day grilling. Members shouted and scoffed. Several times during the session, the usually ice-cool Juneau was taut and trembling as he replied to the questions of committee members who frequently interrupted his answers.

A good part of the attack came from MPs from the hinterland. If U.S. content were reduced on the CBC, and on CTV stations where available, viewers in the large Toronto and Vancouver metropolitan areas, and all of southern Ontario and southwest British Columbia, could get all the American programming they wanted from U.S. channels off-air or through cable. Up the line, though, far from the border in what one angry member of Parliament described as "the captive areas," viewers were out of luck. The backwoods, the cultural mid-Canada corridor from St. John's via Timmins to Prince George, not excluding Edmonton and Calgary, did not take kindly to their access to American programs being limited by some bureaucratic intellectuals in Ottawa while the big cities along the border, not to mention specifically Toronto, luxuriated in American choices.

The broadcasters, moreover, had gone to work on the public, and mail to MPs was heavy, a nice occasion for a bit of self-justified demagoguery. The committee, a couple of weeks earlier, had been only too glad to hear from the CAB and CTV and get ammunition. The powerless backbenchers on the committee could requite their complaining electors and at the same time vent their indignation against the apparently quite powerful CRTC bureaucrats who had not been elected by anybody.

"We have no damned intention of listening to the CBC crap for another four hours," said one MP from an area where only the CBC was available. Juneau would soon be "jamming" American stations and sealing Canada behind an "electronic curtain," said another. And so on. Was Juneau familiar with songs like "Squaws Along the Yukon," as named on a company list of Canadian records? "With this kind of Canadian music, the Canadian people would turn off their sets." Increased Canadian content could mean "two more hours of wrestling between Yukon Eric and somebody from Toronto." Juneau was "tilting at windmills" with a "narrow, chauvinistic, regulatory approach." Would a Canadian pitcher playing with an American baseball team be classed as Canadian content? Did any of the full-time commissioners have a background in private broadcasting, which would qualify them to know what they were talking about? Is there "any real difference between American jokes and Canadian jokes, American humour and Canadian humour, American rock music and Canadian rock music?" The commission's "mandate in a democracy is to do what people want," and what people wanted was American programming.

Juneau patiently answered, sticking to points and avoiding arguments on open-ended questions. Why encourage them? He took pointed exception to charges that the commission was building an electronic curtain ("We cannot avoid feeling as a group insulted by the use of . . . those terms to us") and insisted on respect as public servants trying to do the work legislation requested of them. One of the rare committee members who supported Juneau, gesturing towards other members of the committee, commented acidly that "if Quebec ever separates, Canada won't hold together ten minutes because these guys wouldn't spend a nickel to stay Canadian."

It was a case of the proverbial anthropologist being examined by the apes — which, despite the fearful grunting and groaning, always does wonders for the anthropologist's reputation.

The official announcement on Canadian content by the CRTC was made at a press conference on May 22, 1970. CTV, which originally was to meet the 60 per cent Canadian prime-time requirement by September 1, 1971, was given a thirteen-month extension to October 1, 1972, more than two years away. Of the 40 per cent non-Canadian content, or two hours in prime time (defined as 6:30 P.M.–11:30 P.M.), only three-quarters of it, or ninety minutes, could be American. The 30 per cent Canadian-music rule for radio stations was reaffirmed.

The TV stations were also allowed to calculate their content averages for each calendar quarter rather than every four weeks. The AM radio stations in turn were allowed to count their 30 per cent Canadian-music content over the whole broadcast day rather than in strict four-hour periods. There were other minor concessions. Together they represented a willingness of the commission to meet the commercial broadcasters halfway.

"We believe," Juneau declared at the news conference, "that the prophets of doom, the messengers of mediocrity, will be overwhelmed by the new generation of competent, creative, confident artisans and by all those of preceding generations who have already demonstrated their freshness of mind, their talent and their capacity for inspired leadership."

This declaration would be quoted many times. Patrick Scott commented in the Toronto *Star* that "no challenge could be issued more clearly than that — nor could it be accompanied by a greater willingness on the part of the commission issuing it to make it easier to meet." On the principal objectives, "the commission remains as tough-minded as ever — and Juneau, in fact, seldom has sounded tougher."

Allan Fotheringham, in his regular Vancouver *Sun* column, launched an attack on "that avaricious band of bank balances" — the private broadcasters — for having managed to mislead the public with scare tactics. "All Pierre Juneau and the CRTC are doing is making our greedy private broadcasters conform to a standard of logic they would be forced to adopt if they operated in any other country. . . . The broadcasters are wailing woe and ruin. They've been wailing that ever since the CRTC and its predecessor began to put curbs on their natural desire to fill the airwaves with commercials, interspersed by programs. Juneau . . . is right. I think I can suffer through having at least half my program content created by my own countrymen."

"Inferiority complex ends," read an overline on a background article in the *Sun*, referring to a quote from an ACTRA spokeswoman. "CRTC stands firm against its critics," announced the *Globe and Mail*. "A landmark of Canadian history," said Pierre Berton. A Canadian Press wirephoto of the occasion showed Juneau in close-up from the side, seated, talking, his hands expressively taut. Behind him, facing forward, cheek by jowl in overlapping portrait, was Boyle, holding his pipe to his mouth. They looked like the double portraits of royalty or statesmen sometimes seen on old postage stamps and coins.

In the years following 1970, Juneau and Boyle took the message forward in several public speeches. They became, in a special sense, politicians, leaders of a movement, advocates   in the case of Juneau, who had the mien and style for it, even something of a prophet. Boyle was the pipe-smoking, story-telling street-corner philosopher specializing in folksy evocations of Canada's rich and various human landscape. Juneau was a courtly, knowledgeable and passionate statesman. Both were articulate.

The role seemed to come with the territory. The Fowler Commission of 1957 had pointed out that the broadcasting authority, in making such important decisions, needed to get out and explain to the public what it was doing and why. Juneau and Boyle did. The charges, coming from members of the parliamentary committee and spread abroad by elements of the cable and broadcasting industry, that the CRTC was trying to lower an electronic curtain along the Canadian-U.S. border, for example, could not be ignored, no matter how ill founded and out of context they were. The spring of 1970 had taught them that. These charges were met head-on. Juneau in particular, who liked to deal with large ideas and keep the broad context up front, would offer virtually a wholesale critique of the commercial subjugation of television.

"The commission is arguing," he explained in a speech to the Ottawa Canadian Club, "that unless we do something about it, Canadians are going to have their choices dictated to them by a distribution system which will inevitably find it more economic to pipe all over Canada the overflow of mass-produced American programs rather than support programs that are relevant to Canadians."

The threats to a free TV-viewing choice imposed by commission regulations are less than "the threat of control and censorship by statistical strategies which may very well be developed outside of Canada to support the sale of toothpaste, chewing gum, cigarettes, tires or deodorants.

"How can anyone look at a magazine rack or at the shelves of any bookstore or at the film advertisements of any daily newspaper or at any TV schedules, and then express some ridiculously righteous indignation about the threat of our being starved of American material? The only possible polite comment about such nonsense is: Fiddlesticks!"

Juneau quoted from a recently published book by commission member Northrop Frye, who commented on the effects of "curiously abortive cultural developments" in Canada. Frye said the country had shown a "lack of will to resist its own disintegration." He said Canada "is practically the only country left in the world which is pure colony, colonial in psychology as well as in mercantile economics."

Juneau's speech also had an appeal to action, directed specifically to broadcasters, to throw off their "inferiority complex" about what might be accomplished, to connect with "the richness and reality of this country and this people," to use the abundant creative talent in Canada and add to our choices. The broadcasters had heretofore failed to do so. Canada's ability to survive as an independent nation was in question.

Juneau would explode when accused of censoring the American flow and jamming Canadian culture down people's throats. "That's sick!" he once charged back in an interview. "Anyone who would say that must be obsessed. We're just getting a whisper of our own material into the picture. We're just trying to breathe, for God's sake, in an atmosphere completely dominated by U.S. material."

Juneau and Boyle became even better known public figures than they already were, perceived not as bureaucratic functionaries but as policy-makers and decision-makers, who understood implications and interconnections, who had vision.

3

The Media Get the Message

THE IMPACT OF THE 1970 Canadian content hearing was particularly strong. A full-length feature written by Toronto journalist Jack Batten appeared in the April 1972 issue of *Saturday Night*, a sequel to Fulford's 1970 introduction to the "revolution." It was entitled "The Essential Pierre Juneau, or, What the CRTC Is Up To."

The CRTC, Batten explained, was the "most powerful single body in the Canadian broadcasting business, possessing the authority to make or break fortunes, not to mention cultures." And Pierre Juneau, its chairman, was "some kind of SuperCivilServant." CRTC employees, who gave Batten this message, loved their boss, not entirely a common situation in Ottawa, the reader was told. The commercial television and cable operators, who did not love him, showed their respect. Their approval, Batten found, was "surprising, putting it mildly, since, on its four-year record, the CRTC is as likely to dump on broadcasting's established entrepreneurs . . . as it is to smile on them."

All agreed, however, that Juneau owned "in quite astounding supply the right equipment for his monstrous job." His education in philosophy (postgraduate work in Paris) and his rise through the National Film Board and the Board of Broadcast Governors to the CRTC chairmanship were seen as exactly the impressive kind of credentials for the job. Another part of the "right equipment" was some useful friends. Juneau knew Trudeau back in the *Cité Libre* days. As a young man, he once roomed with Marc Lalonde. And Juneau relied on these friends "to protect his independence from outside pressures." A broadcaster, also a Liberal insider, was quoted as saying he was "convinced that before Juneau took the job at the CRTC, he told Trudeau and the others, okay, he'd be the chairman, but only as long as they guaranteed to keep the politicians off his ass. And they have."

"Score . . . for Juneau," Batten joyfully reminded the reader.

Then there was Juneau's "savvy." He was part of a group of tough intellectuals that was finding power. One Toronto lawyer quoted in the story saw Juneau as "the prototype of the new French-Canadian civil servant" that did not surface in Ottawa until the Trudeau years, one of the "hard-driving activists . . . who are sharp and tough and right on. They're at the top and running things." Another called him "the sharpest cross-examiner for a non-lawyer I've ever met." At commission meetings, Juneau, according to a senior commission officer, wasn't the kind of guy to dictate to the other full-time members, "but he has a way of letting the information generate a natural decision that he favours."

Batten himself was of the decided impression that Juneau had "Balls" with a capital B. Juneau seemed to be, in fact, "Trudeau's one true radical." The Trudeau government "talks about local power, about emphasizing nationalism and favouring new ways over old ways." But it was Juneau who, "more than anyone in Ottawa, acted on those notions."

The CRTC was "against . . . *big, old and other.*" It was "for . . . *small, innovative* [a favourite word around the commission offices], *local* and *Canadian.*" Several CRTC decisions, giving names, companies and cases, were enthusiastically cited to prove the point.

Many problems remained to be solved. The move towards a Canadian broadcasting system was "slow, and, for many Canadians, painful." But "at least the message is clear that the CRTC and Pierre Juneau care about us and our communications."

Credentials, Friends, Savvy, Balls. "SuperCivilServant" was on the job.

Only a few months later, as if by telepathy, the Toronto *Star* carried a large drawing of SuperCivilServant, or "SuperJuneau" as it described him, right across the top of the entertainment section's front page. Maple-leaf belt buckle, CRTC signet ring on the fist, cape flowing, flying in the air over Lake Ontario and the U.S. border — "POW!" — he was delivering a lightning blow to U.S. television signals.

Your average 1972 nationalist was likely tremendously elated at what the Batten article told him, particularly if the nationalist came from southern Ontario and shared the same thought patterns as Batten, Fulford and the Ottawa cultural bureaucracy. If he shared the same general environment as the Toronto lawyers and broadcast operators that were cited, he would not have any inkling of how much the article was an exercise in self-deception.

There was, for a start, something curious about the decisions cited in the article as examples of Juneau's desire for "new ideas and new people" and of the CRTC's "aggressiveness and independence." These decisions had been described in heroic terms. "*Small* and *local* [do take] precedence over *big*," despite the howlings of the big operators, the story went. The CRTC had indeed been "thinking and enforcing." Concentration of ownership was out. But on close examination, the decisions turned out to be mostly housekeeping about which share of the pie different kinds of commercial operators would get, keeping cable and television ownership separate, a straightening out of the rules. And when you straighten out the rules but don't change anything else, the big usually get bigger, given enough time.

In cable, large multiple holdings remained, despite the special opportunity for decentralized ownership created by an order-in-council requiring repatriation of foreign-controlled systems and broadcasting stations. One of the largest of the reorganized cable holding companies, Premier Cablevision — Victoria, Vancouver, Coquitlam and York (Toronto), repatriated from CBS control — acquired an Oakville, Ontario, system and Keeble Cable (Toronto). Agra Industries, until then an agrocorporation involved in rapeseed oil and soft-drink bottling, acquired six different cable systems in smaller centres. Canadian Cablesystems ended up with nine systems acquired from its previous incarnation (Famous Players), including London, Hamilton Co-Axial, Grand River (Kitchener) and Metro Cable (Toronto), as well as a 50 per cent interest in the Kingston cable system.

Maclean-Hunter Cable added Lakehead Video to its multiple cable holdings. A reorganized National Cablevision ended up with systems in Montreal, Quebec City, Sherbrooke, Victoriaville and Cap de la Madelaine, and additional systems or licences for Laval, Rouyn-Noranda, Malartic and Val d'Or.

Standard Broadcasting, controlled by Argus Corporation, and owner of large radio stations in Toronto and Montreal, acquired the Ottawa CTV station, CJOH-TV. Baton Broadcasting (Bassett and Eaton families, CFTO, with John Bassett in charge) added four radio stations and the Saskatoon CTV station. Western Broadcasting added several radio stations and increased its interest in BCTV to 44 per cent; BCTV owned the Victoria as well as the Vancouver CTV stations. Selkirk Holdings — multiple radio, television and cable properties

— picked up CHCH Hamilton and added to its equity in many other holdings. CHUM Limited picked up CKVR-TV Barrie, CJCH-TV Halifax, CJCB-TV Sydney, CKCW-TV Moncton, CKVN radio Vancouver (call letters changed to CFUN), and CFRW-AM and CFRW-FM (now CHIQ-FM) Winnipeg.

These developments all took place during the Juneau regime, either prior to the Batten article or in the years immediately following. Much more serious was the way in which takeovers occurred, with the outgoing licensees free to sell the public licences. This system was left untouched. In one case (the CHUM Winnipeg acquisition), the transfer price was twenty-three times the asset value of the stations. Juneau, whom Fulford fashioned as a revolutionary and Batten called a radical — a *Saturday Night* radical — was not about to interfere with the private appropriation of public property. The double irony was that eliminating the possibility for private transactions would have been simply a classic implementation of public administration principles, radical only in the sense of foreclosing legal profiteering by a few special interests. The door to concentration that the Board of Broadcast Governors had opened was left ajar by its successor, with inevitable consequences far down the line.

One of the glowing examples in Batten's article of the CRTC "thinking and enforcing *Canadian*" was the Canadian-content rules. Another was the ban on the use of microwave to relay U.S. signals to cable systems distant from the border. Except for a passing qualification that there were some loopholes and sliding concessions, these dramatic accomplishments were not discussed. In fact, by April 1972 the commission had drawn back so far in forced retreat on Canadian content in television as to make the 1970 "landmark" hearings almost meaningless, except for the image they projected.

In two stages, culminating with proposals in March 1972, the commission had reduced the scheduled prime-time Canadian requirement for private television from 60 per cent to 50 per cent. At the same time, it had broadened its definition of prime time from 6:30 P.M.–11:30 P.M. to 6 P.M.–12 P.M. Further, the requirement that only three-quarters of the imported programming could be American, in prime time as well as generally, was eliminated. Under the new formula, all of the peak viewing hours of 7 P.M.–10 P.M. could then be filled with American imports, rather than just ninety minutes, or half as originally projected, a wholesale change backwards. Moreover, the basis on which content averages were to be calculated, already adjusted from every four weeks to each calendar quarter, had been further broadened to a whole year. This would allow stations to lever more of the Canadian-content requirement into the slack summer months.

But the proposed, announced, projected, diluted, adjusted and finally emasculated percentage requirement was a fiction anyway - had been all along, even while the noisy 1970 hearing was going on — because Canadian content did not necessarily have to be Canadian in any meaningful sense of the word. Coproductions with a certain level of Canadian participation, usually made in Canada, were allowed to count as Canadian content, although the conception might be American, the stars American and the treatments altogether American. (Coproductions aside, what better way to maintain ratings in colonial Canada than to imitate or graft on dominant American programming fashions and personalities, to feed the colonial strain, accompanied by full Canadian-content credits?)

The coproduction loophole wasn't a secret. It was really not so much a loophole as a well-marked and lighted thoroughfare. Murray Chercover, president of CTV, explained the logistics to everybody who was willing to listen. Financially, there was no other way, he said. Chercover developed a nice, aggressive patter of doubletalk to make the implications go down easily. "There is nothing less Canadian about [coproductions] as a result of orienting them to the foreign market," he once told the parliamentary committee, without blinking. (Mind you, to Chercover, as he had earlier explained to a Vancouver reporter, the Apollo flights to the moon and the Kennedy assassination were Canadian, if you get the point.)

Both Chercover and John Bassett (whose CFTO production arm, Glen-Warren, was handling most of the coproduction work) purred with satisfaction at the success of the coproduction detour. "Makes the CBC look sick, doesn't it?" Bassett boomed, contemplating the frenetic activity in his mammoth Studio One and his expansion plans, which would make CFTO second only in size to NBC's Burbank plant.

Juneau also understood. "We will certainly make sure in our conclusions," he assured the parliamentary committee in 1970, "that coproductions are encouraged and that there are incentives in the rules for coproductions."

Even at that early stage, Chercover's ploy for stealing a Canadian-content base had become notorious. The CRTC's coproduction guidelines, ultimately announced in 1972, simply codified how the game should be conducted in order to count points.

As for the ban on feeding U.S. signals to northern Ontario and the West, distant from the border, via microwave and cable, well, alas, it wasn't. The public announcement that the CRTC would not allow microwave links to cable for importing U.S. stations was made December 3, 1969. The public announcement that the CRTC would allow the microwave links after all was made April 10, 1970. The line held for a mere four months.

The commission had written, in its short-lived December 1969 declaration, that the penetration into Canada of U.S. networks and stations and their advertising markets, through microwave feeds to cable, "would represent the most serious threat to Canadian broadcasting since 1932 before Parliament decided to vote the first Broadcasting Act. ... It could disrupt the Canadian broadcasting system within a few years. The fact that through force of circumstances many U.S. stations now cover other parts of Canada, and that some of them seem to have been established mainly to reach Canadian audiences does not justify a decision of the commission which would further accelerate this process." The Aird Report of 1929 and Canadian broadcasting tradition were invoked. "The commission is determined that the hope and spirit embodied in the Broadcasting Act ... will be successfully achieved." Famous 1969 words.

There had been considerable backlash to the banning of this use of microwave. The 45 per cent of Canadians distant from the border did not take kindly to being denied what the other 55 per cent had been receiving all along. Some population centres, like Edmonton and Calgary, were quite important.

The Canadian-content proposals, which envisaged fewer U.S. programs on Canadian stations, had also just been announced. This only increased the distemper in the hinterland over the prospect that U.S. channels would not be allowed, via microwave, for cable distribution. The April 1970 microwave surrender took place just a few days before the Canadian-content hearings began, conveniently undercutting the tumult in the nick of time.

The idea was that the second service (CTV) should be available and solidly in place commercially before U.S. channels on cable would be allowed. The U.S. channels would then be introduced only gradually. Mind you, the only way to extend CTV and give it that commercial solidity was to dilute its Canadian content requirement and increase its U.S. program allowance, after which the U.S. channels via cable, with still more U.S. programming, could then be added. The retreat was brilliantly won, even if the front was lost.

Pierre Trudeau and Marc Lalonde, the friends in the right places who, according to the Batten fantasy, were part of just the right equipment for SuperCivilServant, were not much help in protecting Juneau's independence from outside pressure in this case. Some of the most unpleasant attacks from backbenchers on the parliamentary committee came from Liberals. The notion that Juneau was "on his own," free from outside pressures, like a ghost-rider in the radio-wave sky, was rather bizarre in the first place. What should have been obvious was that friendship to Trudeau and Lalonde was itself a highly political thing, and that all public people are denizens of a political environment upon which pressures intrude, sometimes directly, sometimes subtly.

By the spring of 1972 the glow was off the 1968 Liberal rose. There was a falling out at a gut level. The acceptance of Quebec power in Ottawa was under siege. The federal election in November 1972, when the Liberals ended up a minority government, was to send shock waves through the Liberal party.

The federal government's own CRTC, high profile, with its edicts about Canadian content and U.S. channels — the appearance that by bureaucratic fiat it was telling people what was good for them — was an open target.

In the spring 1970 session of the parliamentary committee on broadcasting, a Liberal member explained to Juneau that the mail received by the members on the content issue was "almost unprecedented" and that if the members were concerned, Juneau should be able to understand why: They had to be elected every four years. What he didn't say, but what could be easily inferred, was that if too many of them did not get elected the next time, then the government of Juneau's friends and their common outlook would not be re-elected. Then who would protect Juneau's independence? Then what would happen to Juneau's passionate vision of the country? Blaik Kirby, writing in the *Globe and Mail*, mentioned matter-of-factly that the Cabinet had eventually told Juneau to ease up on the microwave question. Whether this was true or not, Juneau did not have to be told. The day-to-day Ottawa circuit consists of people at all levels talking to each other — from ministers to clerks, the parliamentary debates and encounters, the telephone calls, memoranda, meetings, lunches, receptions, games and socializing, first-, second- and third-hand gossip, passing remarks, the morning *Globe and Mail,* the Quebec pipeline. This network radiates signals and sends hidden instructions about its preoccupations and priorities, what can be hazarded, what is acceptable behaviour (what an acceptable "radical" is, for that matter), what must be done to keep things together.

In subtle psychological, cultural and historical ways, Juneau was a prisoner of his friendship with Trudeau, Lalonde and their shared backgrounds. Juneau was not about to embarrass them. The CRTC people and the Liberal insider who had assured *Saturday Night*'s reporter of Juneau's independence from politics couldn't see the political woods for the trees that were part of their own country-side, too.

4

The Commission at Work

THE MAIN FUNCTION of commercial television and cable in Canada was, and remains, to make as much money as possible. The main stated function of the CRTC, on the other hand, was just the opposite: to ensure that television and radio programming in Canada was as Canadian, diverse and original as possible, and not American. At another time, in another place, two groups with such conflicting objectives might have fought it out openly, from opposing trenches. But, with exceptions, this would never happen. How could the commission manipulate relationships in favour of its objectives if it was doing things that continually disrupted the relationships with acrimony and hostility, particularly when the disruptions might work against the commission, since U.S. programs, and hence also the stations and cable systems that delivered them, were extremely popular?

A partial answer is, of course, to break off old relationships and establish new ones — change licensees, deny renewals for cause, oblige stations to fulfil their original promises and let them go under if they could not make it. These tactics would force the government to undertake real structural reform, to take a critique of the structure as well as of its symptoms to the public. There was more than enough rationale for it. An applicant given a licence because of certain promises should rise or fall by those promises. In Great Britain, television licensees were changed in competitive hearings at term. In the United States, Nicholas Johnson, a member of the Federal Communications Commission, was calling commercial television "the principal pusher to a junkie nation," and corporate control of television "perhaps the nation's greatest single tragedy"; Johnson was conducting a personal campaign against the industry, not because he expected to win but because, in the circumstances, there was nothing better to do.

But for Juneau and his vice-chairman and successor, Boyle, and the rest of the commission, as with other ruling groups circumscribed by their own immediate environment, this was impossible. It represented too great a personal departure and too great a departure from their own narrow society and the way politics was played in that society. Juneau's career as a public servant and an associate of the ruling Liberals would be cut off; he would be burning his bridges behind him. But quite likely, the commissioners never even entertained the notion of such a strategy.

The commercial television and cable operators, besides, had audiences and subscribers and some friendly members of Parliament, and also active lobbies, and the cash flow for the requisite lawyers and public-relations men — had, in that sense, the real power of an industry with a large and continuous revenue. No countervailing force of any weight among the public was readily visible. Juneau and Boyle, seeing themselves as independent spirits, might try to keep the lines open to the few random public-interest groups and encourage their participation at hearings, particularly if it suited the commissioners' immediate purposes, but after that, even if those groups knew what they were talking about, what could you do with them, anyway?

Whatever the selfishness of private television and cable owners, however many promises they had broken and no matter the opportunism of their programming schedules, they, their lawyers, their lobbies, the kind of collateral and lines of credit they used, their financial statements, the corporate structures, the faces and the names, how they did things, were a known world. They were the industry. When they appeared at a hearing, collectively or individually, they represented something concrete, something of real magnitude.

The hearing process itself — hour after hour, day after day, year after year, hotel after hotel, volume of transcript after volume — involved a steady, if unconscious, behavioural tuning. Although at some high-profile hearings there might be significant public participation, the vast majority of hearing time, day in and day out, involved licensees and their operations and corporate affairs. For the commissioners, simply to do the job meant speaking in the licensees' language, using their categories, following their modes of thinking. This behavioural tuning worked on the other side, too. The CAB would never again attack the CRTC in the hysterical fashion that they had during the Canadian-content confrontation.

This was the prosaic reality of it. Batten recorded instead Juneau's penchant for springing philosophical questions on unwary supplicants. This idiosyncrasy never meant much with unphilosophical private broadcasters, except that they had to indulge Juneau as best they could. The habit, however, greatly enhanced the SuperImage of Juneau and the commission. Even journalists and lawyers weren't used to dealing with so-called philosophical approaches. Here was something exotic. The main effect of the philosophical touch was to encourage the philosopher in his illusion that the routine was more meaningful than it was.

The 1970 session of the parliamentary committee on broadcasting illustrated that even when the commission and the industry were having a spat, and appeared philosophically indisposed to each other, there were ready interconnections. The committee was itself madcap comedy. One committee member (Ray Perrault, Lib., Burnaby-Seymour) went so far as to explore the possibility of direct CAB representation on the CRTC. This was too much even for the president of the CAB, who thoughtfully volunteered, however, that a recently retired broadcaster could make a significant contribution during the in-camera sessions, to provide the necessary experience and understanding. By the time Juneau appeared, several of the committee were ready to pounce on the chairman for the commission's lack of knowledgeable people from the "private side" of broadcasting.

To their surprise, Juneau rolled out a few connections. Ralph Hart, a special adviser and senior policy officer, for example, had been a prominent member of the advertising industry as director of advertising for a very large firm and as executive director of the Association of Canadian Advertisers (an industry lobby); he was "one of the most experienced men in the buying of time on television and radio and other media." There were also other CRTC people who had been in private broadcasting over the years. Why, commissioner Therrien was a broadcasting engineer who had been associated with private telecommunications and broadcasting, and on the nontechnical side with a Rimouski station.

Vice-chairman Harry Boyle, whom Juneau and others saw as an ex-CBC person, cut in at this point. He had been with a private station, a very small station in Wingham, Ontario, for five years, almost from the day the licence was granted. Knowledge of financial realities, which the committee was so concerned about, was in those days immediate. Boyle's job had "embraced all aspects of the operation of that station from managing it to taking out the garbage."

"I am very happy to hear that," said the requited questioner (Stanley Schumacher, P.C., Palliser) of Boyle's witnessing. Boyle in his own mind might have had disdain for a lot of private stations, but growing up regulatory on Metcalfe Street, he was learning fast.

It seemed, in this little bit of political farce, that members of the parliamentary committee, with scatterbrained righteousness, were trying to push the CRTC into bed with the industry it was supposed to regulate, while the CRTC, in Laurel and Hardy fashion, was busy showing that it was already in the bed. "We're optimistic," Juneau kept saying. The importunate committee members never stopped to wonder if any funny habits might eventually develop under the sheets.

They need not have got so pushy anyway. The work of the CRTC with the private operators was always a family affair. Although the CAB had not been privately consulted prior to the Canadian-content proposals, they had, along with other lobbies *cum* trade associations, met with the commission earlier in the year regarding cable policy. This and other consultations continued under the rubric of a "consultative committee," involving the executive committees of both the CRTC and the CAB. "It was the policy of the BBG before and it is the policy of the commission now to maintain constant consultations with the trade associations," Juneau stated. (Fulford's image of the CRTC as a "revolutionary" departure from the BBG begins to fade.)

"In the normal course of events," Juneau also explained, "we might see several broadcasters in the same day." This was over and above the hearing process and the "consultations." The reorganization of CTV, for example, was undertaken by the network and the CRTC in tandem, at a series of closed working meetings. CTV (as well as the CAB, the Canadian Cable Television Association (CCTA) and the advertising industry) was to have a permanent lobby, in the form of a director of government-and-industry liaison. The aforementioned CRTC senior policy officer and former executive committee member of the advertising lobby was subsequently given responsibility for liaison with the broadcasting industry associations.

"Our relationship with the commission, and the BBG before it, is an intimate one," CTV president Murray Chercover explained at a later hearing. "If there are questions relative to the fulfillment of any particular . . . responsibility, surely those questions have in the past, and can in the future, be addressed in the context of our day-to-day communications and/or informal or even formal committee sessions." Public hearings were largely cosmetic, Chercover seemed to be saying, and why not keep them that way?

Other relationships were intimate, too. Juneau became a good friend of John Bassett's.

"When I was at the CRTC, I got to know most of the CTV board members very well," Juneau himself would offer later. He went to pains to describe publicly, and to reinforce, the closeness of the CRTC's relationship to its licensees. This was an essential part of the strategy: encouragement and a co-operative attitude as a technique for getting things done.

"There are clichés about cooperation between the regulators and those who are regulated," Juneau was to joke at an annual conference of the CAB at the end of his period on the commission, in 1975. "Those clichés come from south of the border. I suggest that we levy an import duty on clichés . . . If people mean by 'cooperation' that we don't believe in constant fight, that is right. . . . My attitude to broadcasters is that they are all nice people. So look them in the face, and don't turn your back on them." The CRTC "needs [the broadcasters'] cooperation," Juneau explained. "I have been too long in the field of creativity to believe that you can get results without a willingness on the part of people to do things."

(Years later, after Juneau and Boyle, Pierre Camu, a president of the CAB and a schoolboy friend of Juneau's, would be appointed to the commission's chairmanship. This raised more than a few eyebrows, even indignation from those with principles, but it was not such a change from the past as it seemed.)

The Juneauesque strategy of co-operation was a wonder all by itself. It didn't need clichés from south of the border. First, the CRTC had to keep virtually all licensees alive, regardless of what the licences were meant for in the first place or what new objectives needed to be added. Anything else would undermine co-operative feelings. When Juneau explained in a general way to the parliamentary committee, "We do not think we are in this responsibility to let that sort of thing [a bankrupt industry] happen," every broadcasting outfit could step jauntily to the bank.

In market situations, businesses regularly fell by the wayside; but except for unusual, marginal cases, the CRTC would bend over backwards to avoid casualties.

Licences were originally awarded for themselves; the operator was not supposed to have any vested interest in the licence. Whether the operator continued with the licence was secondary to the licence's purpose and definition; the operator's only legitimate claim was to receive depreciated book value for his assets. Now, public licences became instead a business guarantee.

Similarly, denying renewals for anything but the most outrageous transgression just was not done. That would be too much like punishment, whereas the approved method was encouragement. The nonrenewal of the licence of a Yarmouth, Nova Scotia, radio station, which had been cited by Fulford as an illustration of the commission's toughness, involved instances of deliberate and unrepentant censorship of the news by the licensee when requested by friends and advertisers. He was virtually asking that his licence be terminated. The CRTC had no other choice in the matter, as indeed commission insiders told a *Financial Post* reporter at the time, to reassure everybody what a freakish exception the Nova Scotia case was.

As for original promises of performance, if the commission insisted on those, every private television licensee in the country would be wiped out, and a good part of the CRTC's own job, too, with considerable political fallout.

Next, the idea of competitive hearings when licences expired (renewal time) or changed hands, and the elimination of trafficking in licences, had to be put clear out of mind. It was better not even to think about these measures, and certainly not to talk publicly about them. When the need to stop the private sale of licences was put to Juneau in this period by Victoria mayor Peter Pollen (who enters this story later on), Juneau sidestepped.

Then, the revenues and profits of licensees had to be protected, particularly in television. If television licensees weren't financially healthy, they could not contribute to Canadian objectives, even if keeping them healthy meant sacrificing the level of Canadian content.

But a regulated rate of return was not enough. Licensees apparently needed to have the psychological security of wide profit margins in order to take the plunge into the kind of changes the CRTC was after, like scheduling Canadian content in the high-revenue peak viewing hours. Co-operation depends on trust and fellow-feeling. The greater the demonstrated profit margin, the more room there was for the CRTC's persuasion and encouragement to go to work. It was a matter of timing. There might be some tough scolding, but in creating the right psychological mood, the CRTC could not be too solicitous in demonstrating that it would look after the financial well-being of its licensees.

Part of this togetherness was keeping the public out. The financial returns of broadcast licensees were kept confidential (and are kept confidential to this day). When the CRTC and CTV, at the 1970 Canadian-content hearing, were openly disputing CTV's financial capacity, based on the profitability of member stations, the public and intervenors had no idea what the figures were. The commission and CTV were not telling, either. Their confrontation at the hearing had the appearance of a bizarre debate where the two sides threw conclusions at each other but omitted any reasoning.

The relaxation in 1972 of the Canadian-content requirements — a historic policy reversal — was based on an independent financial study of CTV commissioned by the CRTC. This study was also kept confidential. Nobody from the public, in a public hearing or out, had a chance to analyze the study and check the CRTC's conclusions.

The ostensible reason for not making financial statements public was to keep the information from competitors. Another purported reason was respect for the confidentiality of someone's business matters. Competitors, however, would have a good idea of who was doing well and who was not. And this was a publicly regulated industry. Because of confidentiality, the public had no clear knowledge of how much of a killing the more prosperous licensees were making. Moreover, for cable licensees, where there was no competition, financial statements were also kept confidential. Intervenors at rate-increase hearings could not see the figures upon which to base their interventions. Mind you, there was no rate-of-return regulation of cable, that is, regulation of subscriber charges based on a percentage of return to the licensee for its investment, like telephone regulation (and there still isn't). Fulford's admiration of the CRTC for having changed cable to the "status of a public utility" in his fantasized "revolution" was nothing more than a nice thought.

Of such is "cooperation between the regulator and those who are regulated" made. The reports of the media, moreover, were colourfully evocative of the CRTC's idea of itself. The commission could imagine itself the passionate champion of the public interest and the nationalist conscience, a considerable cut above the likes of its counterpart in the United States. The private broadcasters could continue to view themselves as the put-upon entrepreneurs, forever at the mercies of a powerful regulatory agency and its philosophical chairman, and succeeding by strength of their own willpower and cleverness despite the obstacles (although the smartest of them realized that the CRTC might be a friend after all).

5

The Great
Colonial Evasion

IRONICALLY, the greater the commission's skills of argument and cross-examination, the greater was the illusion.

Picture the commission at work in its public hearings. It sits on a raised platform. It is physically above supplicants and critics. It looks down on proceedings. It has the intangible aura of power that comes with all officialdom and its ceremonies. Agendas are published. Simultaneous translation is in place.

The chairman controls his hearings. He sets time limits. The commissioners can interrupt for clarification, if so inclined. They can scold if angry, and make wisecracks if they want to deflate pretension or show off.

Only the commission can undertake and follow through on lines of questioning. A licensee or intervenor cannot submit the commission to questioning, or even oblige an answer to a specific question. The commission can follow some lines of questioning and leave others. It can follow a line of questioning for a couple of minutes or for several hours. It can wander far afield or cut to the point. Arguments it does not want to deal with can be terminated simply by not responding to them, or can be sidetracked by diversionary questioning or by paternalism. The more gentlemanly the paternalism, the better it works.

Now add Juneau's formidable skills and his presence, and that philosophical touch. Juneau's mystique could make grown men nervous. Private broadcasters would not usually be as intellectually agile and practised as Juneau. The commission, moreover, handed down the decisions and could make things difficult. Individual broadcasters would habitually be cautious at a hearing, afraid to be combative.

These arenalike encounters were what the public and the media saw. But like all soap opera, it wasn't quite that one-sided. At first, the passionate and articulate Juneau, an unpredictable quantity it seemed, intimidated his regulatory clients, the broadcasters. Then, almost from the beginning, the tables turned.

Inevitably, the broadcasters got used to Juneau and what he wanted to hear. Since licence renewal was virtually assured, it slowly became clear, after the sound and fury, that there was only so much the commission could do. This realization was inevitable, too. The more that the CRTC's tough or impassioned talk was repeated, the less tough and impassioned the commission seemed. In the regulatory to and fro, familiarity breeds equilibrium.

No, the CRTC could not be — but yes it was — the secret son of the discredited Board of Broadcast Governors. Its vanity kept it from recognizing its true parentage. It was going to succeed where the weak-willed and fuzzy-minded BBG had failed. "We are optimistic," Juneau told the parliamentary committee in 1970. "We have accepted the mandate."

These were inspiring words. But have you assumed that the historical function of the CRTC was to achieve Canadian broadcasting objectives? That would be a cardinal mistake. The historical purpose of the CRTC was to allow the Liberal government to avoid facing up to the situation and taking necessary action. The whole CRTC exercise was a masquerade of the system to avoid talking about the obvious: that because of the economics of television, commercial television was inappropriate for Canada, and measures for establishing new public television should be undertaken.

The fine words that Juneau and Boyle used, that LaMarsh and Pearson had used in their Broadcasting Act, and that the Fowler Committee had used, came from the Canadian public-broadcasting tradition. But while they all retained the words of the tradition, and in most cases the sensibilities, and posted the appropriate icons (like the Aird Commission) on the right occasions, they tried doing without the public-financing substance behind the words that had created the tradition and given those words a practical life in the first place. This was a convenient case of bad faith. It was also first-class practical folly.

If, like Juneau, these people had a duty to give regular sermons, they drove themselves into going over the old words with increasing passion and eloquence, words delivered apart from the real life of their origins. The most peculiar part of these passionate declarations by CRTC chairman Juneau was that the exercises might be

performed in front of the private operators, which lent to the pro-
ceedings a quality of the bizarre. What in the world was happening?
"Something goofy is going on here," Allan Fotheringham would
write of an "entreaty" by Juneau to the Canadian Association of
Broadcasters. "To these people, he might as well be speaking in
Chinese."

The 1965 Fowler Committee had indulged in the same evasion.
The report's biting analysis of the performance of the new commercial stations indicated that the creation of these stations had been a
disastrous mistake and that new public television should have been
established. For the Fowler Committee to have come out against the
licensing of these new commercial stations, however, would have
been to condemn the recommendation on that point made by the
Royal Commission on Broadcasting of 1957. This royal commission
had been chaired by none other than Robert Fowler.

The Fowler Committee instead turned to the regulatory agency
as the object of its solicitude; the agency, in the evasion, was the
only thing left to hit upon. The agency should get tough, the report
said. It should have these and these powers and structures.

Juneauesque arabesques now filled the broadcasting-politics
stage, keeping attention off the only issue that counted: the need to
expand Canadian television on the public side.

The CRTC was not primarily a regulatory agency. It was a diversionary agency.

Part II

Public Property/ Private Gain: The Rise of Commercial Television

1

Cutting Deeper
THE LICENSING OF
GLOBAL COMMUNICATIONS

ON JULY 21, 1972, the CRTC announced that it had licensed a new commercial television network in southern Ontario — Global Communications — and that it was inviting applications for new stations in Vancouver, Edmonton and Winnipeg. According to Pierre Juneau, a third national network was a possibility, with Global as the nucleus. In southern Ontario alone, Global would be accessible to six million viewers through six different transmitters.

Here our story properly begins, although one could also say that this is the beginning of the end. The licensing of Global cut another deep wound in the Canadian broadcasting tradition. It was one of the most destructive acts in Canadian broadcasting history.

It established another channel tethered to the American broadcasting industry. It introduced more commercialism against an embattled public-service tradition and opened the door to still further commercial expansion. It hemmed in the already surrounded CBC, left with one channel to support the tradition against not only an Americanized commercial CTV, and now Global, but also the whole weight and power of the American television industry. It diverted valuable resources into the additional purchase of U.S. programs and into the commercial apparatus. It distracted attention from the real, structural dangers to Canadian television and from the only alternatives that made sense. It added to the structural trap instead.

Until the Global announcement, the CRTC had been straightening out the kitchen. The repatriation of licensees from foreign control was a straightforward bureaucratic business, following instructions from an order-in-council. The busy shuffling of the Canadian-content regulations and their loopholes represented just a tidying up after the BBG. On the use of microwave to feed U.S. channels to distant cable systems, the CRTC had capitulated, bringing what was available in the hinterland areas up to the same American mark already enjoyed by cities close to the border.

The licensing of Global, on the other hand, introduced a whole new level of television licences and production, definitively changing the Canadian broadcasting structure.

Before the decision, Canada had two English-language services, the publicly owned CBC and its affiliates and the stations of the private CTV network. The French-language side also had its CBC network and commercial counterpart. There were two exceptions to this two-station structure, CHCH in Hamilton, which had disaffiliated from the CBC, and City-TV in Toronto, a recently licensed UHF station yet to go on the air, whose promoters had sold the concept of a low-powered, local, new kind of television. Excluding available U.S. channels, there was a rough balance between the structure's public and private elements.

The Global decision opened the structure wide on the commercial side. The decision caught almost everybody by surprise. Until just a few days before the announcement, high-placed staff members of the CRTC reportedly had considered the whole Global scheme so much pie in the sky.

"A major factor in the approval of this application," the decision explained, "has been the commitment of Global to concentrate on the development of programs using creative resources of independent Canadian producers and production houses." All of this the commission "strongly hopes ... will contribute significantly to increasing the proportion of Canadian programming in the areas of drama and variety."

The establishment of Global was also intended to help attract viewers away from non-Canadian border stations, particularly the five television stations in the Buffalo area. "Global will help to redress this imbalance and will add diversity to Canadian television services in southern Ontario." In addition, Global would, according to the thinking, repatriate advertising dollars being spent on non-Canadian stations.

Global had undertaken to sell fewer commercial minutes per hour than the total permitted by regulation. The commission put the limit at eight minutes per hour as compared to the twelve minutes permitted other stations.

Meanwhile Al Bruner, the promoter behind Global, was heralded in the press as "Canadian television's sudden superman." What could be more fitting than for a SuperCivilServant to license a superman? Another report described Bruner as "TV's super salesman" and "every inch a televisionary." One columnist, Jim Bawden of the Hamilton *Spectator,* actually subjected the CRTC decision to analysis and left it in tatters, but he could be disregarded; there were too many naysayers in this country anyway.

Bruner, a Leamington boy, the public learned, began his career as a tenor with Detroit radio station WJR when he was fresh out of high school. Some programs featuring him were carried on the CBS network. From WJR he became featured singer with Wayne King, the Waltz King. After the big band era faded, Bruner returned to Detroit and turned to singing commercials. He was singing "See the U.S.A. in your Chevrolet" before Dinah Shore made the phrase a byword.

Back in Leamington, he founded a local radio station, where he began doing Heinz 57 Time, a fifteen-minute weekday show. Radio personality Joel Aldred did the commercials. When Aldred put CFTO in Toronto on the air, with Bruner's help, Bruner, "one of the brightest and hottest advertising men in the business," became his director of marketing. Soon John Bassett squeezed Aldred out of CFTO. Bruner also left. After a detour of a few days in Pittsburgh, seeking his fortune, no doubt at Heinz headquarters, Bruner returned to take over the sales of CHCH in Hamilton, which had just left the CBC network. He helped to make that station one of the richest in the country.

It was at CHCH in 1966 that Bruner conceived the idea of a space-satellite national network for Canada and, when that didn't sell, a scaled-down Ontario version. CHCH gave up on the concept and withdrew its support. Bruner kept pressing until, having been called "everything from a dreamer to a nut, for years," he finally got the licence — the "magic licence," as it came to be called, at least for a while.

"We start to build a program service tomorrow, we're ready to go," Bruner had said on the day of the decision. "We're going to produce one hell of a lot of programming or cause it to be produced, and it's going to be Canadian," he said. Global would provide the resources and broadcast outlets to "turn on the creativity" of Canadian television filmmakers. "The fresh and the exciting creators and communicators will answer this new vista of creativity . . . We are a catalyst," he said in his rapturous way.

The Toronto *Star* told us how Bruner went home the night of the decision "feeling limp after six years of planning and hoping and waiting . . . Joy, the butcher's daughter from Leamington, whom he met in high school there and married in 1945, was waiting."

In his den Bruner had a framed quotation from Machiavelli: "There is nothing more difficult to carry out now, more doubtful of success now, more dangerous to handle, than to initiate a new order of things."

"Yeah," he told a reporter, with a laugh and a big cigar, in the postdecision euphoria, "I've been staring at that damn thing and nodding for six long years." Finally it had paid off.

And begin to build he did. Merrill Lynch, Royal Securities produced a sophisticated thirty-six-page prospectus launching the public stock issue collectively priced at $10 million, representing 57.51 per cent of the company. Bruner (president and chief executive officer) and his chairman of the board, a hotel executive and former stockbroker, crossed the country to discuss with securities analysts and businessmen the upcoming public stock offering. In Vancouver they mentioned plans to extend their service across Canada by satellite and microwave, working closely with new western-Canadian licensees. This proposed third national English-language TV network was envisaged to come on stream early in Global's operations.

The Global "nerve centre" was set for a 93,000-square-foot former steel fabricating plant in north Toronto. Leading architect Raymond Moriyama was hired to turn it into television central. The five-and-a-half-acre plot on which the building sat provided room for expansion if necessary. Even Global's orders of transmitter and studio equipment made the news.

As launch night approached, eighteen months after licensing, the great item of speculation was whether Ontario premier William Davis, or a group called The Young Canadians singing "O Canada," or Al Bruner himself would be first on the screen. Including news, sports and information programs, twenty-five brand-new Canadian series were scheduled, fourteen of which came from independent producers. Full-page newspaper advertisements announced the launching, heralding "Global — your new point of view." On the nonstop four-hour *Launch Special,* on January 6, 1974, Prime Minister Trudeau delivered a taped welcome speech, and Pierre Juneau appeared, as well as Davis.

Less than three months later, Global was facing bankruptcy, and shortly after, Bruner was out on the street.

Why did the CRTC license Global when existing CTV commercial operators had so freely abandoned original promises of performance and become, for all intents and purposes, American stations with their own local and regional (read "Canadian") supplement added?

"A major factor in the approval of this application," to quote the decision again, "has been the commitment of Global to concentrate on the development of programs using creative resources of independent Canadian producers and production houses." These programming plans were to be instrumental in the development of a "Canadian program production industry." Global, moreover, had "indicated that it would be prepared to assist independent producers financially and would work with them in the development of ideas and projects." A minimum of 50 per cent of Global's Canadian programming, Bruner elaborated, would be produced by the Canadian independent production industry. Global, according to its promoters and the decision, was going to be different.

Not mentioned in the decision was CTV's original promise on independent production, circa 1960: The network will "encourage and support, financially where necessary, the work of Canada's independent program producers . . . We would give whatever assistance we could to program idea people in order to broaden as much as possible the base of program supply." Fifty per cent of CTV's program production was envisaged as coming from "other program producers," as distinguished from both the network and affiliated stations. Even the phraseology was interchangeable with Global's.

"He who cannot remember the past is doomed to repeat it," wrote George Santayana. The CRTC remembered the past and repeated it anyway.

One newspaper commentator, reflecting on the rationale of the Global decision, pointed out that "if Global succeeded in putting out popular Canadian shows, CTV could no longer protest its own inability to do so." If the CRTC were in control, however, it should have been able to encourage, intimidate, manipulate or finally oblige CTV to do what it was supposed to have done, without Global around as a lever. If it were not in control, what were its powers to encourage, intimidate, manipulate and oblige the newfangled Global, not to mention little City-TV? Were the original promoters of CTV, like prominent veteran broadcaster Spencer Caldwell, any less sincere in their time than the promoters of Global? It could not be so. The question of Bruner's sincerity would likely be beside the point anyway. More often than not, the original promoters of English-language commercial television in Canada had been replaced at the first bend in the road. John Bassett had eased out the real creator of CFTO, Joel Aldred. Ray Peters, when CHAN-TV (now BCTV) in Vancouver ran into trouble, quickly replaced Art Jones, a photographer who originally got the licence with an inventive presentation and who is now in the dustbin of licence-procurement history. Spencer Caldwell's version of CTV never survived the politicking of the affiliate stations who did not want to surrender any of their territory.

Not everyone in the press swallowed the Global decision easily. Asked by a reporter how Global was going to produce its great Canadian programming on the basis of only eight minutes of commercials an hour, one CRTC member said: "You figure out how they can do it, brother." Another member, Pat Pearce, remarked by way of explanation, "If the man can build a better mousetrap . . ."

A better mousetrap? Of such illuminating metaphors have historic decisions been made in our time.

No matter what the risks and the odds, as long as the CRTC could license Global, they could avoid facing up to their own futility as defenders of Canadian broadcasting objectives.

2

The APBBC
and the Case for
Public Broadcasting

NOW BEGINS THE tale of the conscientious citizens. The Association for Public Broadcasting in British Columbia (APBBC) was incorporated in late 1979 and set out to block the projected licensing of new commercial television stations in western Canada and to put forward a public, noncommercial alternative. It did not take a genius to see that if the Canadian public-broadcasting tradition was going to survive and flourish — and if Canadian broadcasting was going to achieve what the Broadcasting Act and all the fine words since 1932 had said it was meant to achieve — then the structure had to be expanded on the public rather than the commercial side.

In the back of the organizers' minds was the example of the Canadian Radio League which had fought the public-broadcasting battle in the early radio days.

This modest citizens' organization, foolhardy enough to talk about the obvious, had to carry the argument for the whole country. Nobody in all of southern Ontario, where the Global licence hearing was advertised, had opposed the application at the hearing other than on minor qualifications, and all these interventions were from other broadcasters worried about their share of the commercial loot.

The CRTC's proceedings to that point were a good introduction to the exercise. The CRTC had established a licensing round for an entire level of commercial television — a historical watershed and a surrender of the Canadian television structure to the U.S. commercial model — on the basis of an application for a licence in southern Ontario alone. The only notices for the hearing appeared in southern Ontario. Westerners, Maritimers and Quebeckers were not heard from and were not asked to be heard. The western end of the licensing round was fitted in as an afterthought. The first news British Columbia had that a new level of licences was even being considered was the report of the Global decision in the newspapers.

The APBBC put together an executive and an official list of sponsors. The president was myself, vice-president was Vancouver lawyer Bill Stewart and the original secretary was Barbara Gordon, an actress now working out of Toronto. A news conference was held in the Hotel Vancouver. Literature was produced. Sporadic organizing began. A public meeting was arranged in Victoria, and ultimately a Vancouver Island branch was established.

By the time of the licence hearing in June 1973, the APBBC had the official backing of the environmental and consumer movements, elements of the arts community, the B.C. Federation of Labour and the Vancouver and District Labour Council, the B.C. Central Credit Union and local branches of the Committee for an Independent Canada, among other support. There were over four hundred individual paid-up members. The B.C. section of the Consumers' Association of Canada endorsed the association's position. The B.C. Teachers' Federation asked for a postponement of any licence allocation. The original organizational endorsement came from the Canadian Scientific Pollution and Environmental Control Society (SPEC), the province's major environmental organization. The focussing of the issue by the APBBC also prevented several other organizations from being co-opted for "community-support" window-dressing by the most active of the commercial applicants, a syndicate headed by former CBC national newsreader Stanley Burke.

The APBBC also established contact with Edmonton and Winnipeg to encourage similar resistance. There were two reasons: first, to form a broad front covering all three cities where commercial applications would be heard, and second, because one of the major objectives was the creation of a noncommercial network anchored in western Canada by new public stations. This was in contrast to Global's cross-Canada network ambitions and to the CBC and CTV, which were also anchored in Toronto. In Edmonton a small group formed the Edmonton Association for Public Broadcasting, but no major organizing effort was undertaken. The original Winnipeg contact was Heather Robertson, then television columnist for *Maclean's*.

The Winnipeg group decided that it was useless to try to fend off the CRTC's commercial licensing plans, because the decision to allocate a new commercial licence had already been made. Since the commission had asked for the applications, this was not an unrealistic assumption. The group decided instead to organize a community-based application for the licence, as if the "right" people doing the wrong thing, instead of the "wrong" people doing the wrong thing, was all that could be managed. Any commercial operation would be subject to the same financial and programming constraints and would be dependent on carrying commercials.

The Winnipeg group simply was not prepared to take up the necessary public-broadcasting argument because the immediate odds appeared too long. For the APBBC, on the other hand, nothing else was worth doing.

During this period, in March 1973, Juneau spoke to a conference on international communications in Ottawa. His eloquent address was a variation on the theme of participatory democracy. It must be up to society as a whole "to select those media or means of communication that will achieve desired political, social, economic or cultural objectives." Down where Juneau and the CRTC were making decisions, meanwhile, the wonderful choice which Ottawa's participatory democracy was offering in Winnipeg was the wrong thing or the wrong thing, financially structured to reinforce the "inundation from outside the borders of their [region's] experience," which Juneau meanwhile so forcefully deplored in his lofty international address.

The APBBC began making the case: Public broadcasting should be the norm for Canada, not the exception. The licence structure should be expanded on the public, noncommercial side.

The most savage criticism of the American commercialization of television came from within the United States. But there was not too much anybody could do about it other than to create a marginal educational network later to evolve into the Public Broadcasting System (PBS).

The British, who did have a public-broadcasting tradition, and who consequently approached broadcasting from historically different assumptions than the Americans, knew better. Of the three countries — Britain, Canada and the United States — only the British looked at public broadcasting and commercial broadcasting together and decided, on merit, which way television should expand. This seems the only sensible way of doing it. You actually look at the choices and take the best one. But next to what was happening in the United States and Canada, the sensible British proceedings appeared quite extraordinary.

In 1960 the British government established a special committee under the chairmanship of Sir Harry Pilkington to look into the future of broadcasting. The major question it had to decide was whether the British Broadcasting Corporation or the Independent Television Authority, which supervised commercial television operations, should be granted the third television channel. The Pilkington Report had two basic qualities that the Fowler reports and the CRTC could not manage or did not dare: It was truly independent of mind, and it was intellectually honest.

The Pilkington Report calmly took things one step at a time. It took things in the right order. And it proceeded in clear, straightforward reasoning. It began with the purposes of broadcasting, looking at the role and effects of broadcasting whole, rather than just as a commercial chase for the largest possible audience. It then proceeded to step two: an empirical examination of the programming record of the two competing systems. It found that the main offender against the purposes of broadcasting was the commercial network: It lacked balance; the range of programs was not sufficiently wide; it overindulged in triviality and violence, and the relatively few serious programs were only available at inconvenient times. The BBC, by contrast, had demonstrated the structural freedom to strike a balance in its schedule between catering to the mass audience on the one hand and providing variety, challenging viewers, and achieving other social and cultural broadcasting objectives on the other.

To the excuse that in chasing after numbers — even if it meant churning out the same mass fare — the broadcaster was only "giving the public what it wants," the report pointed out just how misleading the phrase was. It had the appearance of an appeal to democratic principle, but the appearance was deceptive.

> [The phrase] is in fact patronizing and arrogant, in that it claims to know what the public is, but defines it as no more than the mass audience; and in that it claims to know what it wants, but limits its choice to the average of experience. In this sense we reject it utterly. If there is a sense in which it could be used, it is this: what the public wants and what it has the right to get is the freedom to choose from the widest possible range of programme matter. Anything less is deprivation.

"Those who say they give the public what it wants begin by underestimating public taste, and end by debauching it," one witness before the committee had put it memorably.

The choice, the Pilkington Report explained, was not between either "giving the public what it wants" or "giving the public what someone thinks is good for it" and nothing else; it was the area of possibility between the two — being "sensitively aware of the public's taste and attitudes" but also, by exploring the range of possible programming, "giving a lead."

"To make good things popular and popular things good," was how a BBC executive described the working objective.

These observations about television were unexceptional. Even commercial television spokesmen in Britain agreed that television should broaden and deepen public taste. U.S. network spokesmen had, from the heyday of radio, made similar observations about public responsibility; statesmanlike oratory went with the territory. In Canada the Fowler reports and later umpteen declarations by Pierre Juneau and Harry Boyle were full of them. Applicants for each and every radio and television licence in Canada were required to explain — item 34 of the application form — their "conception of public-service broadcasting."

The Pilkington Committee, however, then moved to the logical conclusion. It recommended that television be expanded on the public-broadcasting side and that the third channel go to the BBC. It did not consider, in reaching this conclusion, where the money would come from, or the ready availability of financing by commercials as compared to the legislative requirement for raising the BBC licence fee. This was irrelevant to the question, as the committee knew; the public would be paying for the new channel one way or the other. Financing by licence fee made as much sense as, if not more than, raising money through commercials. A recommendation for the appropriate funding mechanism — in this case an increase in the licence fee — was then attached as a matter of course.

The incorrigible committee, which kept stating the obvious, did not stop there. It pointed out that the commercial television service had a "fundamental, constitutional weakness." It had two purposes — first, to provide a television service that would realize as fully as possible the purpose of broadcasting, and second, to provide a service to advertisers and make money. But these two purposes did not coincide. In fact, to a greater or lesser degree, they opposed each other. This fault was "organic." No amount of "specific controls by regulation" would offset it. "It would be difficult to the point of impossibility to force the companies to produce items to fit a programme not suited to their interests as sellers of advertising

time," the report explained. This had nothing to do with particular
policies. It reflected "the real distribution of power . . . behind the
apparent distribution by formal constitution," where the supervi-
sory Independent Television Authority appeared to be in the
driver's seat. Besides, the creative activity essential for a good
broadcasting service could not be generally compelled by the
exercise of regulatory powers.

In Canada the Fowler Committee in this period had made much
the same analysis of the failure of regulation and the contradiction of
purpose of commercial television — and then, in one of those spectacular cases of *Ottawa interruptus,* had avoided the logical inference.

The Pilkington Report of 1960 was tabled in the British House of
Commons in the spring of 1962. The last, intense phase of politicking for the new channel began. The BBC understood only too well
what was at stake: Unless public broadcasting was awarded the third
channel, the general quality and balance of television would deter-
iorate. The public-broadcasting tradition, and the BBC with it, would
be destroyed.

A public-broadcasting system needs to maintain a large presence
with the public for two reasons. Without a large presence, it is not
doing its cultural job, which has to do with society as a whole.
Second, if the proportion of the total television audience watching
the BBC, for example, fell too low, its claim to be the national instrument of broadcasting would be impaired. Its political strength to
defend the scale of its licence fee and possible future expansion
would also be impaired. It would be caught in a diminishing circle.

On the other hand, by doing innovative, demanding, individual-
istic or minority programming as part of its mandate, or even public affairs when entertainment is available on another channel, public broadcasting deliberately chooses a smaller audience. The public
system is thereby faced with the choice of sacrificing its large public
presence, absolutely essential to its purpose, or sacrificing its mandate for individualism and balance in its programming, also absolutely essential to its purpose. In order to maintain a large presence
with viewers, and at the same time to undertake the greatest range
of programming, public broadcasting needs more channels than
those allocated to commercial exploitation. Proliferation of channels should occur on the public-broadcasting side.

The BBC fought hard for its position and was successful: It
acquired the third channel.

What happened was analogous to what had happened to Canadian radio in the 1930s when commercial lobbying was defeated,

and direct, objective analysis prevailed: The measure taken actually produced the desired effect. While Canadian television lost out to Americanization, British television entered an unprecedented, flourishing age and became the envy of world broadcasting. British television, according to journalist Timothy Green in the *Universal Eye,* could be divided into two clear eras, Before Pilkington and After Pilkington.

The BBC, although it now had two channels to commercial broadcasting's one, still aimed to attract in total just half the average viewing audience. To go for less would have been to weaken the presence of public broadcasting as a major communicator and entertainer. To go for more, in the circumstances, would have been to subvert public broadcasting's wide-ranging mandate. The numerical channel advantage of public broadcasting, combined with the BBC's traditional independence and its noncommercial structure, produced room for individualism and risk taking that had no counterpart elsewhere.

"The real achievement of the BBC," a foreign broadcaster told Green at the end of the 1960s, "is that it has been like an icebreaker, always pushing back the barriers."

Compare the Canadian situation prior to the licensing of Global to the British situation in the 1960s. Public broadcasting in Canada was in an impossible position: one national television network for English-speaking Canada against three powerful American commercial networks, with massive budgets, spilling over the border and extended by cable distribution, plus one quasi-American commercial network. The political battle between public television and commercial television in Canada was also a question of whether the country would have sufficient programming of its own or be overwhelmed by American programming. To achieve even a passing balance between public and commercial broadcasting, or to create Canadian television different from American television, there should not have been any commercial television licences allocated in Canada to begin with. If British television, which did not have the problem of American spillover and which limited program imports to 14 per cent of the schedule, needed a Pilkington, then Canadian television circa 1972 needed a double Pilkington to give it enough Canadian kick, as well as to fortify the public-broadcasting blood.

Stopping the licensing of more commercial stations was just as important as creating new public ones. If public television were limited to a single CBC channel, while more and more commercial stations and networks were allowed to occupy the structure, the one public system would be relentlessly pushed into a corner and would become increasingly vulnerable financially and politically. Forcing it to meet all the demands for public service and innovative programming would make it even more of a minority presence. The CBC needed to be a central presence in the structure. New, complementary noncommercial services should be established to broaden the range of programming; they would have the freedom for exploration that the CBC, with its broad national mandate, could not possibly accommodate on its own within one schedule.

The Pilkington Report was a classic restatement of the case for public broadcasting. This central case and the parallel nationalist case in Canada — whose history and argument went back to the 1920s and the eventual establishment of the CBC — lay behind the formation of the Association for Public Broadcasting in British Columbia (APBBC). But there was more to it than that for the association. Once one hurdled the phony ideological block that public broadcasting was an aberrant exception — once one started thinking about expansion on the noncommercial side instead of kowtowing to prejudice — the arguments for public broadcasting multiplied.

There was the need in Canada to break Toronto's stifling hammer lock on network programming — to make the Canadian broadcasting system truly Canadian, rather than an extension of programming administrators in only one part of the country. The APBBC insisted that any third English-language network should have its "centre of gravity" in western Canada. This was only possible in a noncommercial context, as we will see. This reason for a noncommercial rather than a commercial extension of the structure was never mentioned elsewhere. Nobody in Toronto or Ottawa, where all approved "national" measures of this sort were conceived, had yet discovered it, nor were they inclined to do so.

The APBBC also opened up at long last the argument for the abolition of commercials on television — the complete and unabridged argument which, until then, had never been made. The main stumbling block to abolishing commercials was always taken to be financial. According to the conventional wisdom, we needed commercials if we were going to have television, whether we liked the commercials or not; the public funding of the CBC was the burdensome exception that proved the rule.

The political problem facing the public funding of the CBC, trapped in this scheme of things, was that the funding came directly out of our tax money. Ideological critics freely called this funding a "subsidy" and even a "loss," the unspoken assumption being that commercial financing was normal, whereas a public appropriation for broadcasting was abnormal and a loss by definition. Commercial financing, on the other hand, though we ultimately paid for it, seemed to come magically out of somebody's closet. It wasn't "our money" the way tax money was. This helped mightily the prejudice that public financing was somehow expensive and financially undesirable, whereas commercial financing was efficient and logical, the right way to do it if only we didn't need to finance the CBC publicly for Canadian reasons.

This myth was very useful for the Procters and Gambles of the country and for the advertising industry they supported, who profited by it. After the original Canadian radio debate had died down as the 1930s expired, nobody had ever questioned it. As we will see below, the APBBC found, simply by setting the myth aside and looking at cases, that commercial financing was the most inefficient and financially illogical way to fund television, not to mention that viewers had to suffer the commercials as well.

"Canada can extend television west and east across this country . . . if we have the will," trumpeted the visionary Al Bruner at the Global licence hearing in September of 1971, as if the CBC and CTV did not already extend west and east. "But extension must begin here in central Canada and then it must emerge as an energy grid . . . Southern Ontario is the resource area that Canada must look to if this extension is to happen. The resource is here; the momentum is here; and we suggest the involvement is here. Global proposes to serve this area in a manner which will allow it to ultimately serve the national purpose."

Of course. Good old southern Ontario.

The Global vision involved an elaborate transmission and time-sharing scheme, including provincial government participation, whereby Global programming would be carried into every nook and cranny of the country. The Global network idea was everywhere — in the CRTC's decision ("a new national program service"), in the Global prospectus, in public statements by Bruner and his associates, in the press (where extension to the West was taken for granted, an impression aided and abetted by Global). A network agreement was worked out for the licence applicants in the West. In Edmonton and Vancouver, where applicants were to be heard first, all but one, an Edmonton applicant, signed up. Global did not want to own the proposed new stations in the West. It would generously provide these stations with the bulk of the necessary network programming, no doubt to the greater glory of southern Ontario's manifest destiny.

The CRTC obligingly looked on. Both the CBC and CTV networks were anchored in Toronto. The BBG had been wrong-headed enough to create a second network centred in Toronto, duplicating the geographical control and perspective of the first. To add still another network centred in Toronto made no Canadian sense at all.

The APBBC did a geographical breakdown of CBC network production for official prime time, 6 P.M.–12 P.M., for a sample week at the end of March 1973. Of twenty such hours, seventeen, or 85 per cent, originated in Ontario and Quebec. One hour came from the Atlantic provinces, half an hour from the prairies, and an hour and a half from British Columbia. "We're fed up with Toronto programs posing as national programs," Heather Robertson wrote from Winnipeg in her spring 1973 *Maclean's* column, in a biting account of centralization at the CBC. But if the geographical distribution of CBC production was atrocious, the CTV distribution was even worse. For CTV, no prime-time network programming was produced outside of Toronto.

Just in terms of production location, given that breakdown, something in the order of 74 to 94 per cent of prime-time network production in a third system would have had to originate in the West and the Atlantic provinces to give those regions a proper share of activity *on a simple population basis.* If you took into account the fact that although French-speaking Quebec needed to be adequately reflected to the rest of Canada, it need not be so fully represented in English-language production overall, then the role of the West and the Atlantic in third-service network production should accordingly have been even greater. Again, on a simple population basis, network prime-time production from the West and the Atlantic should have equalled production from Ontario and should have surpassed production in Quebec, which had two networks virtually all its own.

Global's stirring network proposal allowed for two and a half to three hours of network input weekly from the "Vancouver-Edmonton area," as they put it (seen from Toronto, it seemed that Vancouver and Edmonton were in the same area). Interpolated on a population basis, this would have worked out to five to six hours for the West and the Atlantic as a whole, from a projected Global network production of twenty-one prime-time hours.

Location of production was just the beginning of the issue. What mattered was who controlled programming and where the key decisions were made. Success in getting CBC network production depended, and still depends, as Heather Robertson wrote, "on a producer's ability to sell a show in Toronto; shows are accepted on the basis of Toronto biases and tastes." All the television network division heads (drama, variety, sports, current affairs, news, children's, arts, music and science, even agriculture and resources) are in Toronto. All of the executive producers of network current-affairs and news shows are in Toronto. The decision-making power and the editorial function at these two crucial levels were then and are now in Toronto, along with lower-level editorial control of network current-affairs and news programming, whose day-to-day and week-to-week decisions of what should be carried and how it should be done establish the picture Canadians get of their country, of what the issues of the day are, of the meaning of events. The farm-out of a production to Vancouver to use its studio space was meaningless except for the local payroll and the unspoken insult. In media, look for the gatekeeper; for the CBC, the Toronto office is the gatekeeper.

Any second network should have offered a choice and broken the monopoly. CTV offered American choice. As for Canadian programming, the second network was a wasteful duplicate of the first. Instead of one national news made in Toronto, we got two national news made in Toronto. Instead of one public-affairs magazine program made in Toronto, we got two public-affairs magazine programs made in Toronto. Instead of *Front Page Challenge* made in Toronto, we got *Headline Hunters* made in Toronto. We got all the variety of producers, national news readers, journalists and story editors changing jobs between CBC and CTV, walking both ways across the street.

Compare the largely stylistic choice between the two networks' national news to a national news produced out of, say, Regina, where the greater geographic distance of the story conference team from Ottawa alone would give the news a different quality and mix, although no less national for all that. True variety comes from working out of different milieus. For southern Ontario viewers, the parochial duplicate network structure wasted choices and impoverished their sense of the country. For the West and the Atlantic, however, it meant cultural subjugation. The "national purpose" Global syndicate, with the "national" agency, the CRTC, saying not an unkind word, proposed more of the same.

The CBC, being publicly owned and financed, always took the most abuse for network centralization. The historical blunder of creating CTV was overlooked.

Here again the APBBC refused to take it all out on the CBC, unlike most of the rest of the country, which kept thinking in terms of a single public network. The one-network mental block included the politically hopeless CBC itself which, faced with the protests from outside southern Ontario, had no defence. Although the APBBC argued pointedly for greater decentralization of CBC network production and a breaking up of the centralist mentality, it was at the same time protective of the existing Canadian public-broadcasting base that CBC network headquarters in Toronto represented. A new public network anchored in the West, on the other hand, would freely take up the opportunity of decentralizing decision-making power over network production. It would at the same time enlarge the pools of production and performing talent outside Toronto, facilitating further decentralization of the CBC. It would provide a leading edge for western and Atlantic production on its own terms, breaking down the paternalism and rationalizations of Toronto-based network power.

The first West German national network (ARD), where jurisdiction over broadcasting belongs to the states (*länder*) rather than to the federal government, consisted of autonomous, self-ruling regions, sharing the financing and network schedule. What the APBBC was proposing was basically the West German structure in chronological reverse, a second public network that was federated and non-commercial to complement the first, also public, network that was centralized, with the difference that the network power of the CBC itself needed to be more decentralized in a country like Canada.

If, on the other hand, a third network were to be established in a commercial rather than a public-broadcasting context, it would inevitably be anchored in Toronto, either as conjured up in Global's southern Ontario fantasies or otherwise. As long as the idea of any new network was arbitrarily limited to a commercial context by dominant southern Ontario, then southern Ontario would indeed be "the resource area that Canada must look to." In such a context, it would not make any difference whether a program was produced in the West with western personnel or not. Because the commercial anchor of such a network would still be in southern Ontario, the pressures to tailor network programming to take advantage of that largest single market would be irresistible.

The same would hold true if there were a group of separate stations rather than a network — if nonlocal programming were shared on a syndicated or pooled basis. "Independent" station is a devious misnomer where the station is not independent of commercial forces; the largest single market rules. Where programming is shared — and much of television is so expensive that it has to be shared — a commercial structure is the greatest, bluntest, most unerring cultural centralizer and blender of them all.

The workings of the Canadian commercial television market are a miniature replica of the North American commercial market, in which phony Canadian "coproductions" made in Canada come out Americanized despite the location of the production on this side of the border. The APBBC copiously documented the workings of the "largest single market" in coproduction and other U.S. sales arrangements, including British-U.S. arrangements, showing how the commercial context worked against independence and distinctiveness in program production. The case histories were already notorious and legion. What was needed, in Canadian program production, was the slow and natural cultural decolonization of western Canada and Atlantic Canada through their own free self-expression, and that was impossible in a commercial context within Canada, just as the cultural independence of Canada was impossible in a continental commercial context.

The other major reason for a noncommercial third service was its freedom from commercials. It was all a matter of looking at the question free of the assumption that commercials were somehow necessary, as alleged by its self-serving perpetrators. One of the most intriguing failures of the social and political process in our society, which also tells us a lot about how power works, is that these unceasing interruptions of a widespread and daily activity are allowed to continue. Sports, entertainment, news, current affairs, drama, comedy and so on are produced separately from commercials. Why not decide to do one and not the other and then spend some time working out the best alternative method for financing it, of which there are quite a few?

Commercials are not only interruptions interspersed in other programming. They *are* programming — twelve minutes of programming segments in every hour in a sold-out commercial schedule, or one-fifth of all programming time for everything. An enormous amount of money and manipulative energy is spent on making these program segments and on their production values. Here is the unremovable flaw in the commercial television structure, including the commercially sponsored time of the CBC: The programming segments called "commercials" are privileged pro-

gramming slots used for systematic propaganda, everything that
broadcasting in western democracies is supposed to be against.

Even the people who inflict the propaganda on the public —
perhaps them most of all — do not like it. A survey by the Financial
Times of Canada, published in February 1973, revealed that 64 per
cent of business executives found advertising to be untruthful or
misleading and 61 per cent agreed it was intrusive and annoying.
Among the latter, television fared the worse, 63 per cent finding
television advertising to be very annoying, 33 per cent finding it
mildly annoying and only 3 per cent finding it not at all annoying.

But the issue goes well beyond broadcasting and annoyance, into
the larger economic and historical questions about the conditioning
function of the propaganda. To describe these programming segments for what they were, the APBBC originated the phrase "vendors' propaganda." Merely describing the segments in this way
moved the argument forward to where it should be. It was one
thing to have to suffer "commercials," a name with a vaguely practical ring to it; but being subject to "vendors' propaganda" . . . that
got one thinking.

The proposed noncommercial service would provide, in place of
commercials, independent consumer and environmental information. This would be an integral part of the overall broadcast operation. Also envisaged was supplementary print information.

In this period in the United States, some public-interest advocates tried to offset the worst of vendors' propaganda by taking
the Fairness Doctrine to its logical conclusion. According to the
Fairness Doctrine, established by the Federal Communications
Commission (FCC), television stations were supposed to provide
for free speech and the presentation of opposing views, an unas-
sailable democratic position. Applied to the vendors' propaganda
programming, this would mean counter-commercials proclaiming
the deficiencies of goods, contradicting openly the original commercials, and carried free under the obligations of the doctrine.
The test cases had to do with toys (misleading claims), large-
engine cars and leaded gasoline (damage to the environment),
phosphate-based detergents (also damage to the environment)
and garbage compactors (antithetical to recycling). This posed a
slight problem for U.S. broadcasting interests, since a system of
closed propaganda, on coming into contact with a body of free
and equal speech, disintegrates, and everything dependent on
that system falls to the ground. The only alternative was to make
this application of the Fairness Doctrine disintegrate, which the
broadcasting lobby, the godfather of free and equal speech, duly
addressed and which the FCC duly executed.

The hypocrisy is staggering: a supposedly democratic medium being financially dependent on a closed and privileged system of propaganda. The most staggering hypocrisies, like slaves in early Greek democracy or human rights clauses in Soviet people's republics, are sometimes institutionally the most impregnable ones. In Canada democratically minded people could appeal to the parliamentary tradition of public broadcasting instead and bypass the hypocrisy.

The biggest and most damaging myth of all was that public broadcasting cost a fortune and was inefficient, while commercial broadcasting was free; that the country could not afford any more public broadcasting, while new commercial television was, as far as expense to the country went, neither here nor there, and so the more the merrier, financially speaking. Commercial broadcasting paid its own way, the line went, whereas public broadcasting had to be paid for.

Nothing could be further from the truth. The public pays for all broadcasting. How it pays is the relevant question. The overheads of financing television by commercials are enormous. It would be hard to dream up a more costly and wasteful way of financing television. The overheads of financing television publicly, on the other hand, are minimal.

The waste in commercial television financing is structural. The biggest part is advertising agency commissions and the cost of producing commercials. The agencies siphon off a charge based on 15 per cent of billings for air time. Production costs are on top of that. Add sales commissions, other associated expenses and the net profits after taxes of commercial licensees. Although it is impossible to isolate all these costs with precision, they work out for television to approximately 63 cents for every $1 of revenue remaining for actual television purposes, or about $400 million in 1982 and rising (the similar waste figure for 1973 was approximately $115 million).

This $400 million plus is money that comes out of our pockets for television, but never gets into television properly speaking — programming, transmission, administration. Instead it goes down the drain of a commercial bureaucracy. No other business has such exorbitant overheads on raising money, except for speculative stock promotions whose notoriety and marginality are well understood.

The public pays through the nose and then some. Advertising agency executives in Toronto wax fat at our financial and social expense while beleaguered Canadian television production of drama, comedy and documentaries is hamstrung for lack of sufficient funds. Production in the West and the Atlantic provinces in particular goes starving. This is as topsy-turvy a misallocation of resources as one will find in any collection of anthropological perversities.

By comparison, the overhead cost of collecting income taxes and other federal revenues, from which most of the CBC's budget comes, is one cent and a fraction for every dollar. The overhead cost of raising money for provincially organized television, like TV Ontario and Radio-Quebec, would be in the same order.

Financing a new program-production sector through the cable subscription mechanism, whose machinery and billing system are already in operation, involves nothing more than a bookkeeping entry — nil incremental overhead, in other words.

Any other public-financing mechanism — a licence fee, a surtax on the sale of television sets, a designated cultural tax on nonessential goods and services — would do the same thing. Dismantle quasi-American commercial television in Canada, raise and reallocate the equivalent resources through noncommercial mechanisms, bypassing the overfed and unnecessary commercial television bureaucracy, and suddenly you have an extra $400 million annually for program production, plus whatever increment would be earned from export sales.

For Canada, noncommercial financing of television is the only method that makes any sense. This country cannot afford the waste of commercial television.

Now the truth is out: The commercial financing of television in Canada has nothing to do with practical financial considerations of efficiency. It has to do with political and social power and its use of myth. The users of vendors' propaganda and their advertising agencies impose on the country the financing of broadcasting they want, at a considerable overhead cost to the economy. They can justify it to themselves in terms of using propaganda on the public and of managing demand for their own advantage. Nobody asks them to justify it in terms of the bottom line of the economy. The extraordinarily wasteful overhead is taken off the country's back. Meanwhile, there are plenty of vocal philosophers around to help out with attacks on the public funding of the CBC.

A financially responsible country would decide, first, how much television of its own it needs, and then decide on the most efficient way of organizing its financing. If it did this, television financed by vendors' propaganda would go out the window.

3

One Commercial Licence After Another

NONCOMMERCIAL OR COMMERCIAL expansion; counterbalancing southern Ontario network power and bias; undesirable and wasteful, cost-exorbitant propaganda — these were the questions that had to be dealt with in any new licensing. All the rest was regulatory distraction. The second and third issues were tied to the first. The question of noncommercial or commercial expansion was the demarcation line that marked off other matters.

In Great Britain the Pilkington Committee, in a roughly similar situation, had conducted a full-scale inquiry into the question for its recommendation on a third service. Everybody there recognized that it was a crucial juncture for British television and that the freest and most clearly reasoned discussion had to take place.

In Canada, even moreso than in Great Britain, the preponderant weight of practical evidence was on the public-television side, not least in the areas that concerned Pierre Juneau and about which he was so eloquent. Public policy discussion and a new open style of decision-making were part of the CRTC's style, too, or at least part of the public image of its style; witness the Canadian-content hearings, although cynics observing the confidentiality of financial information and the retreat on the Canadian-content front might have said otherwise. "For sheer atmospherics," Robert Lewis was to write in *Saturday Night* in 1974, in another one of those breathless Toronto national media inside looks, "the CRTC is also the best, most progressive and involved agency of the federal government."

Back in the real world, far from the approved southern Ontario Pollyanna image of the CRTC, the APBBC was facing a different picture. On the one question that counted, in the structural matter of adding a whole new level of television licences and maybe a third English-language network, the CRTC had held no policy hearing, "progressive and involved" or otherwise. That was that. While Global was being heard and licensed, and other related licensing was set in motion, nothing of the essential matters was ever discussed.

The episode had been dealt with instead entirely in the categories of the local Global application. For lack of a well publicized policy hearing, very few intervenors from southern Ontario showed up at the local licence hearing that did occur. Most of the few were commercial operators, among them CTV, not quite the people to bring up the critical structural questions about commercial television. As for the CBC, it was, as usual, too vacant-minded to think of doing so, and too paralyzed to undertake it even if it had. The real issues, in escapist Ottawa fashion, went missing. It was policy hearings, if part of the CRTC's captive frame of mind, but not necessarily policy hearings. The commission fulfilled its function as if on automatic pilot. The Toronto journalists on the scene did not report the missing issues, either. They did not know the issues existed, since nobody had told them that they were there. The matter was disposed of in undisturbed bureaucratic calm.

Never was a system so well levered to make the wrong choice and to defeat the legislation it was supposed to be implementing. It worked like this: New public broadcasting, simply because it is public, requires legislation in most cases to establish its financing. The licensing agency does not have the necessary taxing powers. Legislation, however, requires political initiative, moving a government, holding off special-interest lobbies, the writing and presentation of a bill, public debate, the marshalling of support by political parties for and against, the parliamentary process. By contrast, where the administrative act handing out licences has been left open, as it has in Canada, although not in Great Britain, new kinds of commercial broadcasting require none of this long, democratic parliamentary process. They just need to get a licence. The licensing agency needs only to hand it out. If the market for selling viewers to advertisers is sufficient, the commercial broadcasting promoters are home free.

The public broadcasting choice may be superior in all ways, including financing method. The licensing agency, acting on its own, will make the inferior choice. No comparison is made. No debate of the two alternatives takes place. When one series of choices requires the democratic process of political debate and legislation and has to win out over alternatives, whereas the other series of choices requires only a quick administrative decision and does not even have to face the alternatives, it is not too doubtful which is going to make hay over the long run.

Canadian broadcasting legislation, going back to 1932, always had this gaping hole through which the original aims of the legislation disappeared. The legislation contained the limits of public broadcasting through control of the financial mechanism. It did not, however, also contain the limits of commercial licences so that expansion on that side would also require, stage by stage as in Britain, parliamentary approval.

A licensing agency could, however, break the leverage. Rather than saddling the country with an inferior structural change, the agency could decline to issue new commercial licences, and explain why, putting the onus on the government and Parliament. It could hold a policy hearing, which in this case it should have done anyway. It could refer the question of a new licensing round to Parliament. If it came to having to protest the inaction of the government, it could also resign. Any individual member of the commission could resign.

Alan Plaunt, cofounder of the Canadian Radio League, had resigned from the overall agency, the board of the CBC, in 1941, when his report on the CBC's own operating structure was not acted on by the board. Eugene Forsey and Guy Hudon had resigned as part-time commissioners of the Board of Broadcast Governors in 1962, when the BBG, in part because of political pressure, equivocated over licensing a CBC television station in Quebec City, which already had a commercial station. Whether or not, in this next phase, the television structure expanded on the noncommercial or the commercial side was critical. Who would take a stand?

With the CRTC, nothing remotely like any of the above scenarios would happen.

The first CRTC hearing in the West, in the new licensing round following the Global decision, took place in Edmonton. Global was specially invited by the CRTC to present its network concept. There was an audible gasp in the hearing ballroom when Global was called, as six smartly dressed men, with military precision, stood up at once and took their positions. Simultaneously, a seventh, looking a bit put-upon but nonetheless resolute, distributed scripted copies of the oral submission to the press table and assorted members of the audience. Al Bruner swung into his presentation, introducing the different members of the team, who came in at different points in the script.

The commission explored the question of network transmission costs frontwards, backwards and sideways, as if it were the only real problem to the Global network proposal. No question of any kind was asked about the strangle hold Global would have over network decision-making. The larger questions — where the anchor of a third network should be; the culturally centralizing effect of commercial structures — weren't put to Global, either. That would be too much like the CRTC's putting incriminating questions to itself for having licensed Global. There was only a soft haze of questioning about these relationships. Question and answer contained the same hidden assumption: the Ontario licensee would have the power to give the go-ahead to programming ideas and the West would be dependent.

Despite the safe questioning by the commission, Global for the first time found its concept under question by the small Edmonton Association for Public Broadcasting and by the APBBC, which also appeared. The leading commercial applicant in Edmonton, moreover, did not want to affiliate with Global.

When the Global platoon reached Vancouver a couple of days later, it held a news conference. Here is the account by Lisa Hobbs, Vancouver *Sun* television columnist:

> No matter how charming people from Toronto are, if you scratch them a little bit it's astounding how quickly they become paternalistic about the West.
>
> Take yesterday's press conference held by Global television in the Hotel Vancouver. . . .
>
> There was Al Bruner, the president himself, a frank and handsome man with the engaging confidence of one whose enterprise is backed to the tune of nearly $19.5 million. Bruner is a farmer's son, is now 50 years old, and was a night-club singer for years before he made it really good with his own record company. He's tough, energetic and knowledgeable.
>
> Then there was vice-president and news director Bill Cunningham, looking as if he'd been yachting, with his white duck trousers and gym shoes and sweater tossed off his shoulders, talking lucidly and brilliantly about the way television news should be handled.
>
> And there was Peter Hill, chairman of the board, described in the handout as "a brilliant, young financial consultant." He is president of the Lord Simcoe Hotel, dearly beloved of Globe reporters during their interminable lunches. Hill, who sports a neat black moustache, looked in his white turtle-neck sweater as if he'd been playing polo.

Then there was Sydney J. Banks, vice-president, a veteran of the Canadian entertainment world. It's Banks' job to spend well the $6 million of programming that Global will purchase annually from independent Canadian producers.

We are now onto the subject of community TV and for the first time the issue of widespread and well-organized local resistance to another commercial station, with Global affiliation, was raised.

In the twinkle of a TV eye, the old eastern paternalism reared itself up to full height. Amiable Bruner kept his cool the longest. . . .

But Banks leaped into the breach with vigor, calling the intervention briefs of groups like the Associaton for Public Broadcasting "an attack on the establishment without an alternative being put forward."

"We could stay in central Canada and do very well," he said, obviously miffed. "This extension is very costly."

But now super-financier Hill was getting terribly cross too. "We'll provide 80 per cent of all programming for the prime time sections and for what? We give 50 per cent of commercial availabilities. We're not taking dollars out of anyone's pocket. We're not Johnny-come-latelys."

By now Bruner was into the breach: "We don't need to have driven ourselves into more grey hairs. We don't need to be here today."

That's precisely what some of the intervention briefs are saying. And if this is how Global really feels about the West — that it is doing us a fat favor by coming and, by God, we'd better be suitably grateful, maybe the interventionists are onto something.

The parochial illusion that a few middle-rung businessmen in a corner of Ontario and their backers were the establishment, to be recognized as such in British Columbia for the purpose of attacking, could only have been imagined by a few middle-rung business men in a corner of Ontario, and their backers, who were too far away from home to know better. There have been real corporate establishments in British Columbia, like the CPR and B.C. Electric, but none that British Columbians have taken lying down. Identifying yourself with "the establishment" out loud at a news conference might be profitable for making an impression on genuflecting Ontarians, but it was a foolish way of asserting a case in populist British Columbia in 1973.

One of the familiar characteristics of Ontario expeditions that impose a Central Canadian ethic on other regions of Canada is their spectacular, self-serving innocence. They do not know that they are doing it, and they can make sincere and morally righteous declarations that they are not doing it while they are. With just a little bit of distance from the Ottawa-Toronto incestuousness, Global appeared to be what in fact it was: a jerry-built, financially wrong, historically cockeyed, American-dependent scheme of a few Toronto businessmen subscribing philosophically to the American way of broadcasting. And they had been given a licence by a few other people whose collective brain was missing its western hemisphere.

Hobbs had the alertness to chronicle not just the Global syndicate's application for the licence but also the opposition:

> The question that is disturbing quite a few people as the [CRTC] hearings draw closer is not so much who should get the licence but whether anyone should get it. Any commercial broadcaster, that is.
>
> For as time has passed it has become painfully clear that while the profits are endless the responsibilities of the broadcaster are virtually nil, all the sincere idealism of the CRTC notwithstanding.
>
> This realization hasn't come from any great activity of the brain. People are simply fed up with hours of poor programming chopped up by vulgar, dehumanizing commercials.
>
> But because there are so many people feeling this way today, something new is emerging on the Canadian scene. Television is rapidly taking on all the dimensions of a first-class political issue and it is this mood that will dominate when the CRTC meets here Monday. . . .
>
> All three commercial applicants have stated their intention of affiliating with Global. . . . For the average Vancouver viewer it isn't going to matter very much which of the three applicants gets the licence as Global of Toronto is going to fill the nightly time slot whoever gets it. . . .
>
> When you get miffed enough you start asking questions that normally would not have occurred. One now being asked by those tired to death of the consumerism fostered by barbaric commercials is: Why create another commercial TV station?

The hearing followed. Allan Fotheringham, then writing a column for the *Sun,* was only slightly sardonic:

> It is amateur turn in the whorehouse in the fake-Versailles surroundings of the Pacific Ballroom of the Hotel Vancouver . . . and the non-paying customers are here thick in numbers, eager to watch. It's financial voyeurism.
>
> All the broadcast men are here and the ad men and the market guys with their With It moustaches and the money men in their striped suits and the CBC executives with their hand computers. . . . There are

millions to be made in this room in the next two days and it's, well, fun to watch. Lord Lambton through a transom wouldn't draw half the crowd. . . .

The stern lords of the CRTC hold the whiphand and it is something to see, all this money coming grovelling and sniffing around their feet, obsequious as all hell, promising tap-dancing shows and ethnic knitting bees on prime time Saturday night in hopes the CRTC will touch them on the shoulder and intone, "Arise, Instant Millionaires." Seeing the mighty grovel has always ranked up there with bear-baiting and so the SRO conditions. Man is cruel.

Neighbourhood Radio, one of the constituent elements of what was later to become Vancouver Co-operative Radio (CFRO), presented a dramatized radio satire, pretaped, about a syndicate chasing after the licence and concocting ways to hoodwink the commission. Neighbourhood Radio suggested any new television station should be co-operatively owned and controlled. Video Inn, a video library organization, presented a live cabaret-style satire of the token public-service programming (ethnic, women's programming and so on) that the applicants were promising. Metro Media talked about the incompatibility of commercial-station operation, geared to advertising revenue, and actual community-initiated and produced programming with editorial control of its own.

The Consumer Action League, one of the supporters of the APBBC, talked about the manipulation of consumers by commercials, as if this manipulation was the function of commercial television, a subject you weren't supposed to talk about at these hearings. It had all been decided ahead of time. Playwright George Ryga (*The Ecstasy of Rita Joe*), who at that time was doing a lot of television work, did not appear, but submitted a blistering written submission on the applications and what the CRTC was doing. The applications had little or no regard for the entertainment industry or the broader issues of Canadian expression, Ryga said, "notwithstanding the thinly disguised pretence that all intent is for the 'service of the community.'

"There is a socio-economic shoddiness about all this that is painfully obvious to those of us involved creatively in public life and entertainment . . . Because licences have in the past been issued to organizations with no stronger credentials than these three is no reason the process should continue."

The APBBC, with its representative backing and with a full panoply of officers in attendance, made the traditional public-broadcasting case based on its careful, detailed formal analysis.

The CRTC, subsequent to these hearings in Edmonton and Vancouver, issued a licence for Edmonton to Edmonton Video Ltd. (which would broadcast under the name CITV). Edmonton Video was the one applicant that had rejected affiliation to Global, though it was quite willing to share programming on better terms. For Vancouver, the commission, in an unprecedented move, denied all three applications. This created quite a stir. Something in the order of $1 million collectively in preparing the applications went down the drain, applications the commission itself had asked for.

Harry Boyle privately described the Vancouver decision as revolutionary.   According to a reliable source, Boyle offered the following explanation as to how the commission managed to make the decision. Sunday, the day before the hearing began, Boyle had procured a car and taken the members of the commission around Vancouver to show them what a different, special place it was. Unlike many a June day in the rain forest, the sun was shining. The psychological ambience that was created in the minds of the commission by this spectacular day in scenic Vancouver was conducive to making a bold decision. If it had been raining that day, the CRTC would have handed out the licence.

Compared to the reasoning behind the licensing of Global, the denial of all the Vancouver applications because the sun had been shining on the day preceding the hearing was absolutely scientific. But as for being revolutionary, this was all in the febrile eye of the Ottawa beholder who was not wont to mention that the decision ran away from the main issue just as the commission had been running away from it all along.

The decision carefully avoided even a mention of the public-broadcasting versus commercial-broadcasting question and its ramifications as regards western Canada, as if nobody had ever brought the matter up at the hearing and as if it were of no relevance anyway. The decision also avoided any reference to information hearings on public-broadcasting options and an associated research program recommended by the APBBC to the commission.

The decision was purportedly based on the inadequacies of the specific applications: They had not sufficiently recognized the potential, or countered the "often-expressed opinion that the views of people living in the area are not now being adequately represented either in the area or to the rest of Canada." Vancouver, the decision explained, is a large and growing city with an assured potential in the cultural, educational and entertainment fields. As a seaport providing access to the Pacific, Vancouver can become "a meeting place where new energies and ideas can develop."

The Vancouver decision was the commission's way of pretending to itself that it was actually doing something in light of the opposition at the hearing and the newspaper cynicism. The trouble was that the participants had intervened on a quite different basis — had intervened, in effect, not against the applicants so much as against the CRTC. If the applications were turned down because of their specific inadequacies, the applicants were at fault for the waste of time and money. If, on the other hand, they were turned down in connection with issues concerning commercial applications in general, then the CRTC was at fault for the waste of time and money not only of three applicants in Vancouver but also five applicants in Edmonton. But this latter alternative conflicted with the First Law of Regulatory Agencies: A regulatory agency cannot find itself at fault.

The leading applicants in Vancouver huffed and puffed in dismay. "We are very disappointed that once again people from the East are telling us what we need in television in B.C.," said one. No doubt he was thinking that he knew Vancouver was a seaport long before the CRTC discovered it on the map and officially confirmed the fact in print. (And what was this business of the seaport being a "meeting place" providing access to the Pacific? Unless you were a longshoreman or in the shipping business, you would make your interconnections at the bar in the airport.) "It is hard to believe that the programming plans put forward by the successful Edmonton applicant were superior in any way to those in the Vancouver application," the disappointed applicant said, not thinking that the Edmonton applicant did not have an exotic seaport to contend with.

As early as the Edmonton and Vancouver hearings in June 1973, some of the Global handwriting was on the wall. In March 1973, following a Global news conference, Blaik Kirby, television critic of the *Globe and Mail*, had written a blunt criticism of Global's plans — "Global boon to advertiser, not viewer" — a copy of which was filed, along with other supporting material, in the APBBC submission. The article was based on the ordinary observation that the centre of the schedule was American, already available off-air or on cable, and that a good chunk of the Canadian content was taken up by additional news.

The breathing space in Vancouver lasted two years, but as for the critical historical issue, it made little difference. The commission's refusal to declare on the public-broadcasting issue and to undertake the information hearings and research eliminated a crucial lever of action. The awarding of the Edmonton licence was another decisive factor. It broke apart the western Canadian front and irreversibly entrenched the Global commercial-licensing round in the West. The APBBC mounted a rearguard action out of dogged principle. One by one, the cities went down.

In December 1973 the CRTC held hearings in Montreal for third-service French-language stations in Montreal and Quebec City. Independently of what had happened in Vancouver, co-operative applications were organized in both cities, somewhat along the lines of CTVO, a co-operative which had been granted a commercial licence (French-language second-service) in Hull.

The Montreal co-operative, at the hearing, declined to speak to its application and asked for a delay. It pointed out that a co-operative that depended on establishing itself in the community faced organizational and financial difficulties that private applicants could ignore and that required considerably more time. It also pointed out that a different kind of television was needed for Montreal, a noncommercial, populist and democratic one. For that, a large-scale debate was required involving the participation of all interested groups, including, in the forefront, the CRTC itself. "The only way of assuring that the airwaves as a public service are treated as such is to structure their use so that they can't be treated otherwise." Anything else was an "exercise in mystification."

The vicious circle of the existing procedure was well understood in Montreal. As Le Devoir put it: "The commission wants to know, before granting a licence, what the licensee is going to do, but the co-operative can't answer without an extensive consultation which in turn doesn't make sense unless the licence is first granted."

The Institut canadien d'éducation des adultes (The Canadian Institute of Adult Education, in Quebec) and a Conseil Québécois des média communautaires (Quebec Council of Community Media) stepped in to back the request for a postponement. The APBBC intervened.

In April 1974 the CRTC granted licences for both Montreal and Quebec City to Télé Inter-Cité. The principal of Télé Inter-Cité was a young lawyer, Franklin Delaney, who had spent a few years with the Board of Broadcast Governors and the CRTC, ending as CRTC secretary, from which he then proceeded to become principal shareholder of a radio-station chain, and then head of the Télé Inter-Cité syndicate.

In May the commission heard applications for Winnipeg. The APBBC intervened. One of the applicants was Communications Winnipeg Co-op Inc., the group that Heather Robertson (since moved to Toronto) had started in 1972. By this time Global had disintegrated and been patched together hurriedly by the CRTC with new owners. The APBBC's submission was a pointedly worded analysis of why the commission had to begin rethinking the premises of what it was doing and their implications for western Canada.

The commission, too compromised by events to pay attention, and not interested anyway, plunged on. In September it granted a licence to CanWest Broadcasting. The leading principal of CanWest, Paul Morton, was also a major shareholder of the reorganized Global.

Also in May the Selkirk Holdings station in Calgary, CFAC-TV, filed a perfunctory application to disaffiliate from the CBC, which was establishing its own station in Calgary. CFAC-TV would thereby become the third-service station in Calgary through the back door. The disaffiliation would in effect be creating a new licence. The APBBC again intervened, calling for competitive hearings as in Edmonton, Vancouver and elsewhere. Nobody else noticed, or realized, what was happening. The point, however, in this out-of-the-way, seemingly innocent disaffiliation hearing went to the heart of the licensing system and civil procedure. Radio frequencies are public property. Nobody had a private, vested right to the use of these frequencies or, in this case, to become automatically the licensee for a television third service without facing competing applications. Moreover, without a request for applications and a competitive hearing, neither the commission nor the public had any way of knowing the alternatives — other commercial possibilities, co-operative ownership or a noncommercial licensee. These alternatives were being arbitrarily precluded.

The CRTC, in its decision allowing the CFAC disaffiliation, ignored the matter.

A year later, in April 1975, another hearing was held in Vancouver for a new commercial television licence. The APBBC, continuing in its masochism, was still around and still intervening. Its coalition had in fact expanded. This time, also, the B.C. Teachers' Federation became involved in its own right. About thirty-five organizations either supported the BCTF's submission or independently endorsed a resolution "opposed to the granting of any more commercial licences for television in the Lower Mainland" and requesting the CRTC to permanently allocate a VHF channel solely for public use. The Vancouver School Board supported the BCTF. Channel 10, the last available VHF channel in the region, was up for grabs in this licensing round.

An ad hoc organization called Publicair was put together in Victoria by Charles Barber (later to become a member of the provincial legislature) to circulate a petition calling for the reservation of Channel 10 for public-broadcasting use. The petition garnered 8300 signatures in just over a week. A smaller, separate petition was circulated by two members of the Department of Social Work at the University of British Columbia. A group of six community media organizations intervened against the granting of a new commercial licence. There were other interventions.

Juneau called it a "fantastic ganging up" against the commercial applicants. (Coherent public participation evidently was a "ganging up" when it interfered with the CRTC's predetermined plans.) Remarkably, the extent of the support for noncommercial alternatives had broadened and intensified since the previous hearing in 1973. Given the enormous cynicism by which the whole CRTC exercise was now viewed, and with all the other major cities knocked off by commercial licence awards, the simple existence of this second round of opposition two years later was striking. It was also clear, from the circumstances and considerable newspaper coverage surrounding the hearing, that the question of a new television licence — whether it should be public or commercial — had become a lively issue. Nobody was being stampeded. The Vancouver *Province,* for example, in a long editorial examining the issue, called for the commission to hold off once more. "It would be as wrong for the CRTC to give the channel to any of them as it was then in 1973."

In July the CRTC issued a licence to Western Approaches Limited, on a UHF channel.

4

Financing New Public Broadcasting

THE CRTC WANTED TO KNOW, from the APBBC, where the money was going to come from to establish noncommercial stations instead of the new commercial stations that the commission was in the process of licensing. This question was put with the confident paternalism that befits the self-image of a prominent statutory agency. It was also a question about politics: Noncommercial funding is available in a variety of ways (general appropriation, licence fees, special sales taxes and corporate taxes) if those in political power are prepared to do something about it. The odds were very long against anything like that happening, however, were they not? Any extra money the federal government had for broadcasting would go to the CBC, and that money would go to extend the CBC's transmission system. There was nothing the commission could do. Harry Boyle, who chaired the Edmonton hearing, delivered a little lecture on the point.

Following the initial postponement of a Vancouver licence allocation, the Association for Public Broadcasting in British Columbia (APBBC) did locate an available way of financing a new sector of television on the noncommercial side and, throughout this train of events, was putting it forward. The commission's short-sighted rationalization of its commercial licensing — where would any non-commercial funding come from, right now, in the immediate future? — no longer could be passed off.

This new noncommercial financing mechanism did not require new legislation. It did not require government funding. It did not require fund-raising campaigns. It was also separate from the CBC. It provided for a regular, assured cash flow. It was an efficient financing mechanism compared to the wasteful financing of television by commercials. There was no trick to it, either. It was right there in front of everybody's eyes, if only somebody, somewhere, had reason to see it.

It was through cable.

The APBBC's "colicensee proposal," as it was called, involved the sharing of cable revenue by two licensees rather than one — the existing licensee, looking after the distribution system, and the public-broadcasting organization in charge of the new service. It would turn cable, the great Americanizing force that Juneau fretted about with such public despair, to Canadian use. The chase — the APBBC chasing the commission — began.

The proposal was mentioned, in general terms, in the Montreal intervention in November 1973. By the time of the Winnipeg intervention in April 1974, the mechanism was outlined, and projected revenue figures for western Canada were cited. A legal opinion confirming the feasibility had been forwarded separately to the commission. The CRTC, in its licensing decision, declined to discuss or even mention the matter.

In the Calgary intervention one month later, the figures, method and rationale of the financing were described at considerable length, and specifically in terms of underwriting a noncommercial third service for Calgary and elsewhere. One of the reasons the APBBC argued for competitive hearings in Calgary was to allow for a noncommercial application using the formula, which would have forced the issue. The CRTC, in its licensing decision, declined to discuss or even mention the matter.

The APBBC had already begun another pursuit, by separate correspondence, to intercept the commission before a second Vancouver hearing was scheduled. It stubbornly pushed the matter through to the executive committee, the full-time, voting members of the commission. It requested, first, agreement on the legal feasibility of the colicensee proposal and, secondly, a public hearing on the proposal and related policy. Aside from the general importance of the matter, a public hearing directly on the subject was the only way to get the matter into the open and pin the commission down. The commission instead killed off the proposal at an in-camera business meeting. The request for a public hearing was refused. No reasons were given. A follow-up request for the reasons for the commission's démarche in the affair was also refused on the grounds that business meetings of the executive committee were confidential.

With the colicensee proposal guillotined by the commission, the APBBC turned to the existing licences as they were. It recommended the allocation of these publicly owned monopoly licences to non-profit, viewer-owned organizations, as existing licence terms expired. These nonprofit licensees would capture all of the lucrative surplus revenue from cable and be able to underwrite the new public-broadcasting service with the minimum conceivable increase in rates. In the more lucrative cases, they would be able to do it with no rate increase other than what might otherwise be incurred. This second approach, recovering the whole profit and tax margin generated by the use of publicly owned cable licences, was even more compelling than the first. It also involved a much-needed reform of the licensing system.

At the Vancouver hearing in the spring of 1975, the APBBC filed a complete and detailed discussion of the proposed transition to non-profit, co-operatively owned cable licensees and of the policy factors behind it. This included an analysis of revenue-generating possibilities which, for the country as a whole, came to $150 million annually down the line, at $2 per cable household per month. British Columbia by itself already had high cable penetration.

The commission conducted the hearing as if the proposal, like the colicensee proposal before it, did not exist. At one point, when an innocent intervenor asked for a licensing delay to allow time for mobilizing, Harry Boyle hypocritically chided him for not having definite plans for a public service after the first two-year wait, omitting to mention or to take into account that there was such a proposal on the table.

When the Vancouver *Sun,* in an interview with Juneau after the hearing, brought up the fact that there was a plan put forward, Juneau replied that a co-operative would find it difficult to sell cable subscriptions without transmitting the same commercial programs it deplored. This was irrelevant to the proposal, which took existing distribution of channels as a given and where the "co-operative" would consist of all cable subscribers together. What in the world was Juneau talking about?

At a hearing in Ottawa on cable policy held shortly after, where the APBBC pursued the matter again, the commission tried to prevent the association from appearing. When, after considerable persistence, it did manage to appear — not much better, in the circumstances, than being excluded — Juneau brought up the red herring of CTVO in Hull, a co-operatively owned commercial television station having serious financial difficulties. Was the APBBC aware of this case, Juneau asked, with gentlemanly condescension. "[Juneau] talks to you," wrote Michel Gratton in *Le Droit* just a couple of weeks earlier, "like a father of a family who smiles before the foolishness of his children."

The APBBC was well aware of the case. CTVO was in one of the weakest commercial markets in the country, the second-station French-language market in Ottawa-Hull. Its noncommercial co-operative ethic conflicted with its commercial situation. Early management problems in dealing with the severe financial limitations and the programming expectations compounded the difficulties. As it was, the CTVO licensee lasted four times as long as the original Global regime, with Canada's richest commercial market to play in. Aside from this, though, it was the CRTC that had licensed CTVO; the APBBC was against licensing new commercial television stations. It regarded their financing by the sale of commercial air time wasteful, impractical and undesirable. Cable, with a regular and assured cash flow, was a different creature altogether. If CTVO's difficulties demonstrated anything at all, it was that the commission was on the wrong track. And if Juneau wanted to make a general comment about the management or economic resources of co-operatives, perhaps he should have mentioned the wheat pools or co-operative retail stores.

The co-operative organizations closest in structure to cable — providing a regular, routine service to mass memberships, and paid for directly — were the credit unions. Cable, in fact, had a considerable extra advantage and financial protection — it was a monopoly, a good reason, by itself, for viewer ownership.

There was also a substantial, though small, subscriber-owned cable system in the country, which could be cited as a model — the Campbell River TV Association — which had been running with great success since 1956 and was putting other systems of its size to shame. The CRTC had avoided the Campbell River model; no CRTC official had ever visited it.

In the spring of 1975, as well, the final stage of organization of cable co-operatives was taking place in the four major cities of Saskatchewan, with political support from the Saskatchewan government and with the participation, on the distribution side, of Saskatchewan Telecommunications (Sasktel), the crown-owned telephone company. So there was also a province-wide example of co-operative cable organization based on a well established infrastructure.

There was already a major example of viewer-owned broadcasting licensees on a national scale, in the Netherlands, where mass-membership viewer organizations, representing in effect all of

Dutch society, were responsible for the preponderant part of television broadcasting, as well as of radio broadcasting, in a non-commercial structure.

Juneau, however, avoided coming to grips with these ideas, ideas that upset the CRTC's established relationships. It was the way any bureaucrat would have done it except with great and much-admired intellectual style that hid from most onlookers what was happening. Juneau, wrote Le Droit's Gratton, "handles ... words and ideas like an acrobat without rival; he doesn't need anyone to defend him." Misapplied references, like the one to CTVO, were thrown up to defend the commission's mentality the way a zone defence is applied in basketball. In the mind of the regulatory beholder, they kept any challenge to the CRTC-clientele relationship at bay.

The CRTC, in the Vancouver licensing decision that followed, declined to discuss or even mention the APBBC's proposal or the key policy issues underlying it. The proposal had the misfortune to conflict with the Second Law of Regulatory Agencies: A regulatory agency does not reform itself.

All this, going back to the licensing of Global, was how commercial television was handed the crucial third service, despite every practical and historical reason to the contrary, and how the beleaguered Canadian television-broadcasting structure was given the coup de grace. It was done in the name of the Broadcasting Act and of Canadian identity.

The turn of events in Vancouver in 1973 — the putting over of any licence award — meant that the APBBC, established for a six-month exercise in public education, was kept in existence by its members. This put a public-broadcasting advocate in the field, which led to another discovery. As long as the CRTC was left undisturbed, it had the appearance of being something of a public-broadcasting agency. Juneau's philosophical and rhetorical bent contributed particularly to this impression. Saturday Night's mythology about Juneau and the CRTC came from this rhetoric. If, however, anybody tried to pursue an issue from a broad public-broadcasting point of view, as the APBBC did, the real nature of the agency rose to the surface.

Part III

The Agency That Refused to Regulate, and Other Phenomena

1

Advocacy Advertising
CORPORATE PROPAGANDA

IN SEPTEMBER 1973 the Association for Public Broadcasting in British Columbia (APBBC) filed a brief with the CRTC objecting to the misuse of the commercial format by Imperial Oil to broadcast "political propaganda." The particular propaganda segments, skillfully built around real, visually interesting situations and ordinary people, were carried on *Hockey Night in Canada.*

Advertising agencies and some of their clients had been getting bolder and bolder in using commercial time for corporate propaganda purposes, as distinct from the advertising of goods and services for which the format was meant. The CRTC was like a guard dog in a deep sleep; it could be ignored. The segments produced by Imperial Oil for use during the 1972–73 hockey season were particularly contentious, dealing with investment, job creation, the environment, injecting dollars into the economy, underwriting student employment, Arctic exploration and energy security, at a time when the whole oil industry, in which Imperial Oil was the leader, was coming under increasing debate. More than a few viewers across the country, when the Imperial Oil items appeared, sat up and asked themselves what was going on.

This usurping of the airwaves for special-interest propaganda could not have been more serious. It deeply offended democratic principle. It undermined the very cornerstone of Canadian broadcasting tradition: the principle, in the words of Leonard Brockington, the CBC's first chairman, in 1939, that "there should in general be no preference for any Canadian over his fellow Canadian. . . . Above all there should be no preferences for wealth. Freedom of speech is not for sale at fifty dollars a minute on the air."

This principle was first and foremost a *parliamentary* principle, one of representative democratic expression in broadcasting as opposed to privileged leverage over it by special interests or by wealthy segments of society. The abuse by Imperial Oil and others made a mockery of the fundamental idea, in law and tradition, that the airwaves were owned by the public and not by advertising agencies and their clients. The root policy had been spelled out by Brockington in 1938 and 1939, when the CBC was the regulatory agency. Brockington's defence of principle was direct and classic in its reasoning.

"The free interchange of opinion is one of the safeguards of our democracy," Brockington said, "and we believe we would be false to our trust as custodians of part of the public domain if we did not resist external control and any attempt to place a free air under the domination of the power of wealth."

And, he added under renewed attacks, "I believe ... that our policy is the only one that can be supported because it is the only one that will preserve the essential parts of our inherited freedom ... I cannot escape the conclusion that under our constitution either all of us have an equal right to speak over the air or none of us has any right to speak over the air."

The principles Brockington enunciated were restated in a white paper issued by the CBC's board of governors. A revised version of the paper was adopted by the Board of Broadcast Governors as Circular 51 in 1961, a few years after it took over general regulatory responsibilities from the CBC. Circular 51 was inherited by the CRTC and formed part of the commission's policy guidelines, although it did not have the force of law:

The policy of the Commission with regard to controversial broadcasting is based on the following principles:
1. The air belongs to the people, who are entitled to hear the principal points of view on all questions of importance.
2. The air must not fall under the control of any individual or groups influenced by reason of their wealth or special position.
3. The right to answer is inherent in the doctrine of free speech.
4. The full interchange of opinion is one of the principal safeguards of free institutions.

In the view of the Commission, these principles are not promoted by the sale of network time to individuals or commercial concerns for broadcasts of their opinions or propaganda. The principles can be furthered by the provision of free time to competent speakers to present, without let or hindrance, varying points of view on questions of the day. The best safeguard of freedom of discussion is a policy which permits opportunity for the expression of varying points of view.

The Imperial Oil abuse violated the commission's own policy. It also violated Section 3(d) of the Broadcasting Act that called for "reasonable, balanced opportunity for the expression of differing views on matters of public concern." Who ran the oil industry, how it was being run, for whose benefit and what it did or did not do for the economy were undeniably matters of public concern.

All of this was awkward business for the backers of the propaganda such as Imperial Oil — that is, it might have been awkward if the CRTC had not had its mind on other things, like trying to get more commercial revenue into broadcasting. Nobody at the CRTC paid any attention to what was happening, or if they paid attention and understood it, nobody did anything about it.

Imperial Oil was circumventing the policy not by buying big blocks of time, but by using commercial time slots in repetition. This was probably more effective than a block of time in any case, since the propaganda segments were short, were put together with expensive production values and were slipped into hockey telecasts while the audience was enjoying itself and in an open frame of mind.

The APBBC's brief to the CRTC on the subject of advocacy advertising focussed on two cases, both of which contained misleading statements and both of which were on controversial subjects. The first was a segment about construction of a new refinery in Edmonton:

I suppose you want to know what a job/employment multiplier is ... [This is the construction of] a refinery in Edmonton ... It is? ... It will cost $200 million but that's not what's important ... It isn't? ... What's important is that it will take some 2000 persons to build it ... 2000? ... And thousands more will be working to provide materials and services ... They will? ... That means thousands of more jobs for people who aren't employees of Imperial but are working because of Imperial. That's called ... An employment multiplier! ... It is?

What this slick piece of propaganda failed to mention was that along with the construction of the new refinery in Edmonton, other refineries in Winnipeg, Regina and Calgary would be phased out, resulting in less full-time employment and presumably greater Imperial Oil profits. Even the construction figure of 2000 persons referred only to the peak period of construction. Refinery employment in the unlucky cities would be sharply cut back (in Winnipeg, from 156 to 27). Both the Manitoba and Saskatchewan governments had protested the closings as bad regional economics. The Saskatchewan Federation of Labour had called for government intervention. At the much ballyhooed Western Economic Opportunities Conference in Calgary in the summer of 1973, some employees of the Calgary refinery had picketed outside the conference hall, protesting the shutdown of the plant.

The other propaganda item cited in the brief dealt with Arctic exploration and the wonderful environmental research Imperial was doing to ensure "that man can live and produce in harmony with this truly incredible Arctic world." The item ignored the whole controversial nature of Arctic exploration and of associated developments like pipelines, including calls by several important organizations for the exploration to be discontinued or held up until appropriate research programs could be launched and completed. It also neglected to mention that a couple of years earlier, Imperial Oil was instructed by federal agencies to suspend certain operations in the Mackenzie Delta because of failure to comply with conservation rules.

These items were just two of an extensive series of image-building propaganda segments. There was something out of kilter about a country allowing a foreign-controlled organization to use privileged propaganda to form that country's public opinion, although brainwashing by a domestic organization was no less wrong. The democratic integrity of Canada — its ability to decide wisely and independently — was being undermined. The APBBC called for the provision of air time and production funds by the two networks, CBC and CTV, for counter-items in connection with the two items cited. It also requested a public hearing into the general question, preferably in Vancouver, and asked that in the interim all such political propaganda be stopped. In addition, the APBBC recommended an inquiry into the control of *Hockey Night in Canada* by MacLaren Advertising through its subsidiary, the Canadian Sports Network.

This last request was put forward because broadcasting licensees were supposed to be responsible for their programming, a point written specifically into the Broadcasting Act. *Hockey Night in Canada,* however, was controlled by the Canadian Sports Network, which held the rights, sold the commercial time, hired the commentators and produced the feature material. It also fired commentators when team owners gave the word. The program was tagged by one journalist as "Sponsor's Night in Canada."

The APBBC based its submission on propaganda examples cited from the end of the 1972–73 season. Similar propaganda items continued the following season. Where an abuse is current and continuing, a regulatory agency of any integrity can be expected to act quickly, putting at least an interim halt to the abuse and, if questions persist, holding a hearing to clear the air. The APBBC, having done its duty as good citizens, waited. It had a lot to learn — it is still waiting.

In September 1973, when the submission was filed, an acknowledgement was received. After that, nothing happened. A follow-up letter in November went unanswered. A telephone call in December produced no results except to locate the submission in the CRTC's legal department. By the beginning of April the hockey playoffs were imminent. The APBBC asked its legal counsel what could be done. The lawyer thereupon requested assurance from the commission, to be received within two days, that a response would be forthcoming in no less than six days. As if by magic, after a futile wait of six and a half months, a return telegram arrived within the forty-eight-hour deadline. It said: "The commission is not convinced that it is desirable at this time to hold a public hearing into the general question . . ."

"Not . . . at this time" is one of the most brilliant all-purpose phrases ever invented for use by a regulatory agency. The CRTC pleaded extreme complexity and the need for further study, but the real problem with the issue was its classic, inescapable simplicity. It allowed no room for the commission to avoid standing up to the advertising industry, influential corporate advertisers and commercial broadcasters. The commission refused in the interim to instruct the television networks to discontinue the abuse. It also refused to instruct or request them to provide equivalent time on *Hockey Night in Canada* for counter-items. It made no mention of advertising agency control of *Hockey Night in Canada*.

The CRTC referred self-indulgently to a couple of measures it had taken in other areas with regard to advertising which, however, had nothing to do with the issue it was asked to deal with. Juneau and the rest of the gang at 100 Metcalfe, CRTC headquarters in Ottawa, were no Leonard Brockington. With a routine bureaucratic evasion, a vital democratic principle was betrayed.

The APBBC then filed telegrams with the CBC and CTV to get their responses on the record. There was little doubt about what the nature of the responses would be. "We are certain that you are aware of the fact that all commercials aired by any broadcaster in

Canada, whether from the private sector or the public sector, must be approved by the Canadian Radio-Television Commission," wrote CTV.

The CBC's policy in the books was identical to the BBG/CRTC circular, including the instruction against "the sale of network time to individuals or commercial concerns for broadcasts of their opinions or propaganda." The CBC supervisor of commercial acceptance objected that this policy, which the APBBC had lifted from the CBC's manual and quoted, was "not in the context of commercial acceptance and was, in fact, taken from our policy on party political broadcasts." When asked for a photocopy of the policy on which commercial practice was based, the same man at the CBC blithely sent the very section that the APBBC had quoted from and that he had objected to!

The CBC's Commercial Acceptance directive was explicit enough, too: "Under CBC policy, time is *not* made available for sale on CBC-owned stations for controversial or opinion broadcasting." The CBC's supervisor of commercial acceptance simply denied that the Imperial Oil segments had any controversial or opinion material in them. The request for counter-items was similarly denied. So much for the declaration, in the policy manual, that "the air belongs to the people, who are entitled to hear the principal points of view on all questions of importance," or that "the right to answer is inherent in the doctrine of free speech."

In the United States, without a public broadcasting tradition and with a history of corporate strong-arming of broadcasting policy, two of the three national networks, CBS and ABC, had nevertheless consistently refused to accept "idea advertising" from companies or others. The president of CBS, Arthur R. Taylor, had put the matter much as Brockington had in the 1930s: "To permit this time to be purchased for propaganda purposes, other than certain political views [presumably by political parties and candidates], would mean that those with the most money would get to talk the loudest."

Meanwhile, the media in Canada, with very few exceptions, ignored the story, although similar stories of oil-company propaganda campaigns out of the United States were recognized in Canada as certified news. Morris Wolfe, television columnist of *Saturday Night,* was one of the Canadian exceptions. His analysis of the Canadian story, which detailed the case and the CRTC's evasion, appeared under the headline "The CRTC has rules: why aren't they enforced?"

There was no legal recourse against the CRTC. You could oblige the commission by law to answer, but not to do anything. Without pressure of publicity on the CRTC, nothing was going to happen. Nothing happened.

By contrast, when an issue of balance in broadcasting offended fashionable Ottawa mores — and in those days, respect for Quebeckers and the French language was a new mark of civilized sensibility, Ottawa style, and not bad for upwardly mobile careers, either — the CRTC magically sprang into action.

On September 3, 1975, radio station CFCF-AM in Montreal, as part of its regular programming, began a petition campaign among upset anglophones against the Official Language Act of the Province of Quebec, commonly known as Bill 22. Complaints about the vehemence and one-sidedness of the campaign began to filter back to the CRTC, although unlike with *Hockey Night in Canada,* the station's spokesmen did invite comment from both sides of the issue, especially during the open-line programming. All told, the commission eventually received "several" written complaints and "approximately one hundred" telephone calls. On September 8, *five days after the campaign began,* the commission asked CFCF for tapes of its programming relating to Bill 22. Only four months later, after a detailed tabulation and analysis of the tapes for the two weeks of the campaign, the commission made a major announcement carefully discussing the case and calling CFCF to a public hearing on the matter. A year later, the commission issued a lengthy announcement critical of the station and exploring the general issue of balance. The commission even attached to the announcement a copy of BBG Circular 51 on "Political and Controversial Broadcasting Policies," the same circular they were assiduously avoiding when it came to the Imperial Oil segments.

At the same time, beginning in the fall of 1975, the newly formed Public Petroleum Association of Canada (PPAC) raised the question of Imperial Oil's abuse of the commercial format again. It had been more than two years since the APBBC had filed its brief. By the time the PPAC got into the action, the largely foreign-owned oil industry in Canada was coming under growing examination and criticism. The privileged Imperial Oil propaganda allowed the company to bypass this critical information and present solely its own repackaging of reality in front of *Hockey Night in Canada*'s huge audience.

One of the more notorious items was the portrayal of profits on the basis of each dollar of sales: six cents, or 6 per cent. Quoting profit as a percentage of sales was an old but venerable scam, used off and on in the past by companies large and small to pretend that they were making little money. The figure did not mean anything.

The appropriate figure for rate of profit, which the company, incidentally, used in reporting to its shareholders, was the return on shareholders' equity, or, according to the last financial statement (1974), 21 per cent.

Another item dealt with the need by Imperial Oil for more money (that is, higher prices) if it was going to be able to continue "the big tough, expensive job" of developing oil supply and lessening Canadian oil dependence on other countries. (There was no mention of Imperial Oil's historic advocacy of the export of crude oil to the United States.) It would be difficult to imagine anything that was more obviously a public issue than oil pricing when the price of oil was under government control and when, in the spring of 1976, the issue was exercising the whole country.

The PPAC undertook two lines of action. The first was to produce a "commercial" of its own for use on *Hockey Night in Canada*, in order to reach the same audience. The one-minute film was produced at a cost of only $2,000 with the volunteer help of professional filmmakers. The next step was to write to Frank Selke, director of marketing and promotion for the Canadian Sports Network. Selke claimed that the show was completely sold out. Time could only be resold, he said, at the request of the various advertisers (in this case Imperial Oil). He referred the PPAC to the CBC, CTV and Global for the purchase of another time slot. (A few months later, the manager of sales development for CBC-TV was reported in the *Globe and Mail* as saying that *Hockey Night in Canada* had not been completely sold out for a number of years, even with Imperial Oil as a sponsor. The PPAC patiently brought this discrepancy between the Selke version and the CBC version to the CRTC's attention.)

The PPAC then wrote to CBC president Al Johnson, repeating its requests. "Moreover," it wrote, "the PPAC has no intention of itself approaching Imperial Oil to request one minute of its commercial time. We believe the responsibility for ensuring balance in the views expressed in commercial messages rests with the CBC, not with Imperial Oil." A follow-up letter, pressing for a response, pointed out that "Imperial Oil is continuing its practice of presenting highly controversial political messages in place of product advertising. These messages outline the company's views on Canadian petroleum policy. . . . We believe that the CBC must take ultimate responsibility for ensuring that a balance of views is allowed in matters of political controversy. This week, the nation's energy ministers are meeting in Ottawa to debate the price of oil. Imperial Oil is continuing to influence this crucial public debate through its ads."

Jim Laxer, who headed the PPAC, thought they had an open-and-shut case. They did. "You would have to be out of your mind," Laxer reflected later, to claim that Imperial Oil's campaign was not meant to influence public opinion. Johnson did just that:

> ... Advertising messages of an institutional nature (image advertising) for such advertisers may present factual information about the advertiser's own projects and plans. Such advertising must not, however, urge CBC audiences to adopt a particular attitude or course of action in relation to these projects and plans. In our review of these messages we have found no attempt on the part of Imperial Oil to influence viewers' opinions in a manner which would contravene our Commercial Acceptance Policy.

Now, two years after the APBBC had raised the "advocacy advertising" issue, *As It Happens* duly recognized that there was a story there for a national audience after all. Both the PPAC and *As It Happens* operated out of Toronto. Let's call the program *As It Happens In Toronto Two Years Later. Take Thirty, The National* and others also did stories. This conferring of journalistic value on the issue helped Ottawa journalists, in turn, to recognize that there was a story. The Canadian Press in Ottawa interviewed CRTC chairman Harry Boyle about it. This helped Boyle in turn to recognize that there was an issue there. Confronted by the reporter, Boyle issued grave but noncommittal expressions of concern.

In its submission to the CRTC — its second line of action — the PPAC, like the APBBC before it, dutifully documented the politically controversial aspects of the Imperial Oil propaganda, explaining the obvious. The PPAC also patiently documented, with citations, how the use of Imperial's advertising to influence public policy was taken for granted by the business community. "Al Johnson and his commercial acceptance department," remarked PPAC spokesman Jim Laxer later, "are the only people in the country who do not believe that Imperial's new commercials are examples of advocacy advertising."

But the PPAC hadn't seen anything yet. Two months after the *Hockey Night in Canada* refusal, CFCF television in Montreal, which was "privileged to have several oil companies as our clients," also offered to sell time to the PPAC for the "commercial which you have prepared on the subject of the Canadian Oil Industry." Although the PPAC wanted to reach the same audiences that Imperial Oil had, it accepted the CFCF offer and asked for a minute on the *Late Show* program for the relatively small cost of $280. The PPAC was then instructed to submit its film to the Telecaster Committee, the commercial-acceptance committee of the private television stations. The Telecaster Committee replied: "It is the policy of the Telecaster Committee to approve commercials which advertise goods and services and corporate good will. It is the opinion of the Telecaster Committee that advocacy advertising does not fall into any of these categories and therefore your commercial has been rejected on this basis."

Who could beat that? Meanwhile, CTV and its member stations, including CFCF, continued to sell time for politically motivated propaganda more to its liking. The corporate world and advertising industry, as the PPAC also patiently documented, continued to refer to this material as "advocacy advertising." The broadcasting industry also referred to it as "advocacy advertising."

The account of these events, together with the documentation and analysis, were also filed with the CRTC. As well, the PPAC submission explored at some length the principle of the right to reply on matters of public concern established in the Broadcasting Act and, tellingly, in public announcements by the CRTC itself. Documentation of the issue up to this time also formed a chapter in *The Big Tough Expensive Job*, a book by the PPAC that appeared later in 1976. We know by now that the agency in a situation like this avoids doing anything meaningful while giving the appearance of earnestly doing something. Chairman Boyle, in several speeches following the PPAC submission, promised that the commission would be planning a public hearing, though when, he did not say. As in 1973, the propaganda abuse continued in the interim.

Spring turned into summer. Summer turned into fall and another hockey season. Fall turned into winter. Custody of the issue was taken over by a newly appointed vice-chairman of the commission, Jean Fortier, incidentally a former president of the Liberal Party of Canada (Quebec).

In January 1977 Fortier made a major speech on advocacy advertising, pointing out its implications, in order "to get a public discussion going ... to provoke and maintain an interesting level of discussion." The speech was delivered to a dinner meeting of the Club de la Publicité de Québec (the Advertising Club of Quebec). This was like making a speech on enforcement of the Criminal Code to a conference of gangsters. "I should state at the outset," Fortier reassured the club, "that the commission has not and is not making any plans for regulation."

One of the major points of the speech was that advocacy advertising could create a backlash. A former advertising executive had written in *Marketing* magazine that the risks in advocacy advertising were enormous, Fortier reported, and that agencies should advise their clients not to indulge. (If only the advertising agencies cooperated on this, then the problem would go away, and the CRTC wouldn't have to do anything!) Fortier did not go into the question of why the advertising industry could be expected to advise clients in this way, when it might mean less business for the industry.

The commission, during 1976, had been working on the case of Montreal radio station CFCF's imbalanced treatment of Bill 22 and the Quebec language issue. It decided to establish a subcommittee to study the question of balance in broadcasting in a broad context. This subcommittee would also cover advocacy advertising. It was called the Task Force on Freedom of Broadcast Information, a decisive-sounding name. With a broader context the work of the task force could be stretched out for several years — if the jockeying inside the commission steered it that way — before any firm recommendations were made. Nevertheless, in February 1977 the commission, in connection with its conclusions on the CFCF case, announced that a hearing on balance in broadcasting would take place in the fall. This announcement was made the better part of a year after the PPAC submission. The scheduled time of the hearing was another half a year down the line.

Winter turned into spring. A dinner seminar on advocacy advertising was held at the Faculty of Law, University of Toronto. The CRTC task force was one of the cosponsors, along with the communications law sections of the Canadian Bar Association and the Faculty of Law. By this time new specimens of advocacy advertising had appeared that went beyond the Imperial Oil campaign.

A lawyer for the Association of Canadian Advertisers, in the meantime, had frankly defined the purpose of advocacy advertising as an attempt "to inform and persuade the public about matters that are not directly related to the sale of a product or a service. It reacts and retaliates against unfair attacks upon such of the free enterprise system as remains. It attempts to influence government and the public at election time." Possible countervailing ideological uses of the vehicle did not come into it.

Two of the most ideologically and politically loaded of the new propaganda specimens were by Noranda and the Insurance Bureau of Canada. One Noranda advertisement was a general defence of corporate profit. The "Let's Free Enterprise" series, by the Insurance Bureau of Canada, was aimed at fighting government intervention in the economy in general, and publicly owned automobile insurance in particular. Like the Imperial Oil campaign, the Noranda and Insurance Bureau of Canada entries had magically received the blessing of the Telecaster Committee.

Meanwhile, in the Solarium of Falconer Hall at the Faculty of Law, University of Toronto, the seminar on advocacy advertising was taking place. Jean Fortier was the chairman. Three of the four panel members were from the broadcasting, corporate and advertising coalition; the other was Jim Laxer of the PPAC. A number of "resource persons" from the advertising industry were also made available by the organizers in case anybody needed extra enlightenment on that score.

No CRTC notice was issued to the public or to the CRTC's mailing list; it was not a public hearing. The invitation list included about thirty-five people representing broadcasters, large corporate advertisers and the advertising industry, twenty people from a variety of public-interest groups and the trade union movement, as well as the local and "national" Toronto media and assorted others. A general invitation also went out to the Communications Law Program and to the Bar Association's Media and Communications Law Section, the main sponsor. With the exception of Laxer, invited in his PPAC capacity, no political scientists, historians or sociologists were put on the panel or invited to the get-together. These were the "resource persons" an unrigged seminar would have first consulted about democratic practice, its subversion, the uses of propaganda and the abuses of power. Aside from the panel (representatives from CTV, Noranda, advertising agency Spitzer Mills and Bates, and the PPAC) and a few lawyers, the only people who participated in the discussion were from the Insurance Bureau of Canada, advertising agencies, the PPAC, Imperial Oil and CTV.

Because of the locale, invitations were limited to southern Ontario, including Ottawa-Hull, with three exceptions: CFCF television in Montreal, and representatives of Greenpeace and Co-op Radio in Vancouver; presumably the latter two were to drop in to the 6 P.M. dinner meeting after work. The APBBC, which had first raised the issue and filed an extensive brief, was not notified.

The discussion got down to business. "The law asks us to supervise and regulate," Fortier said at one point, "but we have always said that it is the broadcaster's responsibility to establish the basic rules for his own broadcasting station." In other words, why don't you guys do something so that we don't have to do what the law asks us to do? "It may well be that society will have to accept the phenomenon [of television advocacy advertising] and find ways to adjust to it." Why society might have to accept the phenomenon and adjust to it when the CRTC could save society the abuse simply by doing its duty, Fortier did not say. It didn't take much guessing, though, to figure it out. "The industry, the CRTC, the broadcasters, the advertising agencies, should together develop a philosophy allowing advocacy advertising to take place in a well ordered way, that is, in the event that it has to happen ... Everyone agrees tonight that the adjustment should not be toward a 'fairness doctrine' but some other solution. What is this other solution if it is not a total ban?" The other solution — there was only one other solution — was to keep the talk going until everybody had run out of breath and had decided that the seminar was over.

Seminar: Advocacy Advertising, an edited transcript of the Solarium get-together, was published shortly after, under the white matte cardboard cover of important CRTC documents, complete with the bilingual Canadian Radio-Television and Telecommunications Commission imprimatur and the Government of Canada's red maple leaf logo. The lengthy invitation list was included as a guest list. Anybody could get this significant-looking document for the asking.

The fall of 1977 came and went. There was no hearing.

Years passed. That was it.

2

The Children's Broadcast Institute

A PUBLIC-RELATIONS STUNT

THE LITTLE-KNOWN STORY of the CRTC and the Children's Broadcast Institute is as fine an example of a captive agency collaborating with its industry as any cynic could ask for. That the agency was seen on the outside as "progressive and involved" — the public-agency equivalent of saintliness — makes the example all the finer.

In March 1973 James McGrath, the Progressive Conservative member of Parliament for St. John's East, managed the improbable. Almost two years after his first try, he got unanimous consent from the House of Commons to send a private member's bill to committee for study, a bill that actually threatened a few quite vocal and well-heeled commercial interests. The bill would ban totally all television advertising on children's programming. "At that time," said McGrath later, "we could have pushed through a ban, given the mood of the House and the mood of the country."

"Children," McGrath explained during his campaign, "lack the experience and judgement necessary to deal with advertising. And yet we allow our children to be bombarded with seductive and high-pressure sales methods, produced with subtle advertising techniques which create false images and illusions that can do irreparable damage to their concept of a real world. Many parents, including myself, find it objectionable and unfair to see their living rooms inundated with extremely compelling advertisements that are aimed indirectly at the parent — the real consumer — by way of the child."

The aim of getting the bill before the committee was first of all to give the matter full public discussion. In the process, though, McGrath had generated strong public support. *Saturday Night* editor Robert Fulford, in his Notebook column, speaking as a father, backed up McGrath. Fulford did not try to hide his anger at what the advertising industry and its clients were doing. Children's advertising, he wrote, was "one of the truly grotesque scabs on North American capitalism." He suggested readers write their MPs. The Toronto *Star*, Canada's largest paper, was sympathetic to McGrath's bill. The Consumers' Association of Canada, the Citizens' Committee on Children (an Ottawa-based organization devoted to promoting the interests of children) and all other submissions made by public groups and parents' representatives supported the bill. Constituency mail in favour of the bill was heavy. "It appeared initially to be very much a motherhood issue," said the Liberal MP for Thunder Bay, Keith Penner, who supported bringing the bill before committee.

The cereal and toy manufacturers were not exactly happy about all this. Toy manufacturers appearing at the committee hearing estimated that sales were multiplied ten times by television advertisements. They shouted doom and gloom. The Reliable Toy Co. Ltd., however, was strongly opposed to television advertising aimed at children, and it disputed the view that the industry would be hurt without television. "If there were no television advertising," a spokesman said, "as many toys would be sold by Canadian manufacturers but the distribution of sales from company to company would be different because only a handful advertise on television." One of the favourite arguments of the toy industry, not without its little bit of blackmail, was that a ban on television advertising directed to children would result in the decline of research and development of new toys. It was a favourite argument, that is, until a member of the parliamentary committee established, at least in the case of one of the majors, Irwin Toy, that the company depended on "American ingenuity." Another favourite argument was that the commercials prepared young people for advertising as they grew older. Commercials, in other words, were a kindergarten for cynicism. Most people thought this was one of the damning arguments *against* advertising directed to children rather than for it. It was also a dandy argument for getting rid of television commercials altogether. The toy industry might as well have been recommending a charitable foundation for the promotion of pickpockets and pursesnatchers to prey on small children and thereby teach them how to look after their money. "The more exposed they are to the commercial world, the sooner they will mature and be able to look after themselves in later life," said the executive vice-president of

Irwin Toy Ltd. "The child is ... a heck of a lot smarter with respect to toys than his parents."

A splendid argument was put to the parliamentary committee by the advertising agency representative who handled Kellogg's account in Canada. He was arguing against directing the advertising to the whole family, particularly to the mothers who did most of the purchasing, rather than exclusively to the children. "Children's advertisements talk to them in their own language," he said. "Mothers cannot have the same influence on a child. That's why we advertise directly to children." This short-circuiting of the mother, like the argument for the promotion of cynicism, was, however, one of the key arguments for abolishing advertising. MP McGrath could not have put the words in the man's mouth any more tellingly.

The sugar-coated, junk-food hard sell to children by cereal, candy, chocolate bar and soft-drink companies was also being rapped at this time in both Canada and the United States. Dentists, doctors and nutritionists were getting into the act. Meanwhile, the Kellogg Company of Canada was telling the House of Commons committee that its advertising directed to children was to get children to eat a proper breakfast. Under further questioning, the president of Kellogg admitted that the primary motive of the advertising was to sell the product.

The commercial-broadcasting lobby also went to work. The Canadian Association of Broadcasters (CAB) warned its members about McGrath's bill and invited them to "sit down with your MP ... and convey to him your views concerning this proposal." Some of them did, and suddenly McGrath's bill was no longer a "motherhood" issue.

Where was the CRTC in all this? It was trying to fend off the McGrath bill. Juneau's dialectical performance on this score was especially fine, since he was arguing in favour of something he detested, or seemed to detest. "There is little doubt that television advertising has been used to influence children in a manner that is totally unacceptable," he asserted.

The first CRTC manoeuvre was the "advertising code" diversion. When McGrath first put forward his bill in 1971, the CAB had drawn up a code on children's advertising in an attempt to deflect criticism. When the bill went to committee in 1973, Juneau announced the commission's intention to supervise the code and press all stations to conform strictly to it. Juneau offered that "the new code seems to meet most of the criticisms that have been levelled at children's advertising except the basic criticism that it was totally wrong and there shouldn't be any." Yes, well, that was the point of the bill: that there shouldn't be any. And that is what most of the criticism of the advertising was about.

The code subsequently came under detailed criticism: It had been concocted by the industry, not by the CRTC and the public, the charge was made; it was full of loopholes; there were no penalties for infractions. The political function of a code like this, though, is what really mattered to the industry. However much it was criticized as a code, and was tightened up or watered down, the premise of it would give a patina of legitimacy to an inherently undesirable activity. "Do it," the code said, in effect. "Just be gentle."

Juneau's support of the code, the press duly reported, "gave the advertisers and broadcasters a boost."

Advertisers and their agencies, Juneau told the parliamentary committee, "have much to answer for. Nobody denies that, not even the advertisers themselves. We need no scientific proof for the obvious abuses." The CRTC, however, could not on its own eliminate the advertising. "These hearings have proved," said Juneau's text, "that we still require much proof before the long range effect of television advertising to children on the child's development, and on family interaction, can be established . . . At this time [the commission] is unable . . . to take a sweeping stand as if this was a black-and-white moral issue." Juneau's statement to the committee sounded like the arguments of the tobacco industry in defence of smoking.

If the bill was vigorously adopted, Juneau continued, and could be made effective despite many problems, what would be the effect? The amount of programming specifically for children would definitely be reduced. Revenue of $10 million would be lost to the embattled Canadian broadcasting system. One could seek other forms of financial support, but one was talking about fairly large sums of money. Canadian children's programming was already having a difficult time competing against exploitive American children's programming and its merchandising strategies.

Juneau, in his masterfully paternalistic way, was telling McGrath and the others that they were simple-minded "black-and-white" moralists. He wanted the members of the committee, accordingly, to see all the "circumstances" whereby they could make a "wise and prudent decision." It was the commission's "obligation" as the "public trustee of broadcasting in Canada" to inform the committee of every matter of importance to the system. He proceeded to take them on a short dialectical tour through the broadcasting-policy countryside. He got as far as what he metaphorically called "the polar ice cap [which] may be gradually but surely descending upon us," namely that "to a large extent our broadcasting system has become involved in the entire North American merchandising mechanism." Television advertising to children was just one aspect of that mechanism. This was the "fundamental problem that should be our main concern." It had been brought about by Parliament having decided over the years that our broadcasting system should involve commercial support as well as public funding.

So far, so good. The next step, in logical progression, was to discuss the ways and means of removing television from the commercial support that was the fundamental problem. This step in the dialectical tour of broadcasting policy, however, the commission and its chairman could not take. Their vaunted independence of mind did not go that far, as we have already seen. "Nobody suggests that this [commercial element] should be eliminated," Juneau said. This casual speaking for everybody in the whole country was just a bit presumptuous and not altogether scientific. It was also wildly inaccurate, irrelevant to the objective logic of the situation, intellectually evasive, and pandering. It was, however, also very Ottawa. And it did not offend the CRTC's clients, the commercial licensees. Juneau then wound his way through a long and sundry list of possibilities and commitments, ending up where he had begun, as an apologist. For someone opposed to "the impact of Madison Avenue on the Canadian broadcasting system," it was the policy equivalent of Napoleon's retreat from Moscow.

The Commons committee did not proceed with the McGrath bill. It instead produced a compromise report recommending that the CRTC "pass regulations which would provide that advertising must not be directed exclusively to children." This would allow advertising to remain on children's programming and provide commercial revenue, but it could not directly aim at children and their vulnerability. Juneau, however, refused to act on the report, although the commission did have the authority to make the necessary regulation. Policies adopted by the commission affecting important aspects of broadcasting, said Juneau, had always had a clear reference to objectives established in the legislation (Canadian-content regulations were an example). Recommendations of a committee report lacked sufficient weight. The wording of the report also was not binding, Juneau claimed. The commission's position was that it should not undertake "a very radical change in broadcasting" without specific legislation in the form of an amendment to the Broadcasting Act. McGrath's bill would have been such an amendment.

McGrath was furious. In fact, Juneau was too artful by half. There was, for example, no clear reference among the objectives of the Broadcasting Act regarding the creation of a whole new level of television licences, the possible establishment of a third national network, the pushing over of the structure onto the commercial side by that licensing round, deletion and substitution of commercials on cable, concentration versus diversification of ownership, or television-radio-cable-newspaper cross-ownership. These were questions that really should have been dealt with by Parliament and legislation. Yet Juneau and the other members of the commission had had no trouble straightening away their consciences to deal with these on their own. Wasn't it the unspoken premise of the Broadcasting Act that through the CRTC the government could avoid the broadcasting can of worms and, at the same time, Juneau and other government appointees constituting the commission could make policy without anything so messy as having to deal with politicians?

A few months later, the commission ordered the CBC to dispense with commercials directed to children and did it without benefit of an amendment to the legislation and without any reported pangs of conscience.

About this time the Gallup poll published results of a survey showing that almost twice as many people (57 per cent as opposed to 33 per cent) favoured a ban of broadcast advertising aimed exclusively at children. By this time, too, the financial figures bandied about were coming under question. The Canadian Advertising Advisory Board reportedly claimed that $50 to $75 million would be lost by such a move. This must have been a reference to all advertising on programs that children might watch, including adult programs. This board representing professional manipulators might well start with the highest figure that could be concocted, no matter what it meant. Juneau had put the figure for advertising on children's programming at $10 million. The CBC's figure for itself was $1.5 to $2.5 million, and the CAB's figure was $3 to $5 million, or a total of $4.5 to $7.5 million, considerably short of Juneau's figure. It was understood, however, that the CBC would be dropping children's commercials in exchange for a corresponding increase in its public appropriation. That left only the CAB figure, which was revised downward by the CAB itself to $1 or $2 million, or 0.5 to 1 per cent of total revenues. Most of that, the CAB claimed, went back into the production of children's programs. The Citizens' Committee on Children asked sardonically if the figure had been adjusted downward because the CAB was loath to show what a fantastic gap there was between its revenue and its programming expenditures.

John Bassett, in an unguarded moment in 1973, in rejecting any ban on children's advertising, had allowed that his station, CFTO, would get back in advertising twice what it spent that year on children's programs. "You've got to cover profits and business interests," he said. In that year the rate of return of all private television stations on net assets was 41 per cent before tax, a stratospheric level for a regulated, licensed activity under public protection and, at least theoretically, under licence not to produce profits but to provide a public service. The estimated rate of return of the five largest CTV stations (Montreal, Ottawa, Toronto, Edmonton and Vancouver) for 1973 was an extraordinary 49 per cent. (Bell Canada's rate of return in the mid-1970s was 13 to 14 per cent.) On the assumption that 16 per cent was a respectable rate of return at the time for a regulated, licensed public-service activity, private television in Canada in 1973 made excess profits of $20 million. At an 18 per cent rate of return, excess profits would be $18 million. This was more than enough to replace the $1 or $2 million if all advertising on children's programming was eliminated, or an even lesser sum if only advertising directed to children was eliminated. This would not have been a "very radical change" in broadcasting, just a bit of old-fashioned public administration with backbone.

McGrath reintroduced his bill. It never got to committee. Two Liberal backbenchers talked it out and killed it. The first was a former CRTC staffer, Monique Bégin. The second was a former commercial-television-station news announcer, Mark Raines.

Enter Gérard Pelletier, minister of communications, who allowed to the press that he had supplied the backbenchers with notes. "I'm against advertising directed to children," Pelletier said, as a preface to stating why it should continue. Never in the history of human affairs, it seemed, did so many people neatly oppose what they defended, or defend what they opposed. "The only thing we can do in practical terms is restrain children's advertising," Pelletier explained. "We can't ban it."

It was as if Pelletier had stopped thinking for himself and did not know he was talking just plain silliness. The revenue of private television left plenty of margin to support children's programs free of advertising, and then some. If licensees tried to blackmail the CRTC by pushing children's programs off the air, the commission could reallocate the licences on expiry to applicants willing to meet the licence requirements. Or, through legislation, which Pelletier as minister had the power to introduce, blocks of time on existing channels could be reserved for children's programming and allocated to separate licensees. This programming, like children's programming on the CBC, could be financed by public appropriation. Alternatively, a special tax on excess profits generated by use of the public's monopoly broadcast licences could be implemented, as it had been in Great Britain, and used for children's programming. Or a special tax on television advertising generally, of a few percentage points, could have been used.

A few million dollars annually for extra children's television, where billions were spent on educational systems across the country, could be easily argued against all comers. The public interest and support generated by McGrath's campaign had provided Pelletier with a golden opportunity, the best part of which was that McGrath was a Conservative. But Juneau and Pelletier between them would not implement the reform one way or the other and would not allow it to happen.

"The CRTC is a regulatory body but it does not like to regulate," commented the Citizens Committee on Children sarcastically. However, the commission's paternalistic relationship with its licensees was as amicable as ever, undamaged by any black and white moral action.

In the spring of 1973, just after McGrath succeeded in getting his bill to the committee stage, an "industry group" spearheaded by the toy manufacturers began to formulate the guidelines for a Foundation for Children's Programming. The group presented a brief to the committee on how important it was to have good children's programming and the necessary revenue to produce it. When Juneau, before the committee, defended advertising to children and in the same context said, "I think we all would like to increase the quality and quantity of Canadian programming — especially for children — not to reduce it," bells went off. The industry group sent a letter to the CRTC suggesting that forces be joined. The Children's Broadcast Institute was formed.

Its stated purpose was "improving the quality and increasing the amount of broadcast programming directed to children so that television and radio can play a more creative and worthwhile role in the lives of Canadian children." Its unstated purpose, for the industry, was to deflect criticism from advertising directed to children. A children's programming foundation was perfect. The opportunism of the toy industry, cereal companies and advertising agencies, and the rationalizations of Juneau and the CRTC, were a matching fit.

A series of consultations was undertaken by the research branch of the CRTC. The original head office of the institute, until permanent space could be established, was a room at the CRTC. The funding came from the toy and cereal manufacturers. The first chairman and the original driving force behind the project was Boyd Browne, president of Mattel Canada Ltd. The honorary president was Pierre Juneau. The task force that put together the charter was largely an industry group. The original full board of directors as of October 1974 was a coalition dominated by the advertising industry, toy manufacturers, commercial broadcasters and friends (fifteen of twenty-one, not including the man from the CRTC). The constitution was arranged so that the "private sector," the "broadcast sector" and a "public sector" would each have five members on the board, and another seven would be directors at large. Between the "private sector" and the "broadcast sector" and a good share of the directors at large coming from the same two sectors, control of the Children's Broadcast Institute (CBI) was in safe hands.

The second chairman of the board, taking over from Browne a couple of years later, was Mac Irwin of Irwin Toy, whose testimony to the House of Commons committee in 1973 about the child psychology of the toy industry was so enlightening. The CBI hired an executive secretary and proceeded on the usual rounds of conducting surveys, making contacts, participating in meetings and conducting workshops and conferences. Children's program producers, writers and performers, child psychologists and others were called on to participate. One of the events at which the CBI was represented was the National Symposium on Children's Television convened in Washington, D.C., by Action for Children's Television (ACT). The group was one of the most vigorous critics of commercials directed to children, but the CBI could not touch the subject or it would be spanked. The main thing was that a Children's Broadcast Institute was around and could be referred to in political and public-relations ploys when the matter of children's advertising arose.

The full-page editorial in the *Broadcaster* in June 1974 did not fail to put this tactic to good use.

"While numerous critics, led by vociferous MP James McGrath, have been generating much heat (and little light) about broadcast *advertising* to children," intoned the *Broadcaster*, "another group has been quietly working away in the right area: children's *programming*." It then went through the history of the formation and aims of the new Children's Broadcast Institute, all of which proved, at the end, that James McGrath was the same misguided troublemaker that he was at the beginning of the editorial: "At a time when it is so easy to criticize, this group has taken on the much more difficult task of being constructive. *Broadcaster* applauds all those participating in the Children's Broadcast Institute and urges all members of the industry to support this undertaking. We will be pleased to forward your membership applications (and we'll let you know if James McGrath joins)."

"Why is it unique?" asked an early outline of the CBI. "Because it involves members from the private, broadcast and public sectors of the community. For the first time in the history of broadcasting in Canada, private and public resources are in one organization."

But what happened to the "right area" of concern, children's programming, about which the *Broadcaster* editorial so pompously pontificated, and for which commercials directed at children were supposed to be so essential? "With the exception of the CBC and TVO's Galaxy package," a reporter for the same magazine wrote in 1980 — a typical year — after an uninspiring Children's Broadcast Institute conference, "the fall schedule for children doesn't look much better than it ever has." Neither the CBC nor TVO carried commercials with their children's programming.

Why should commercial operators weaken the punch and profitability of their schedules and lower their net income by added expenditures and air time on children's programming, when it was easier — and financially painless — to concoct a Children's Broadcast Institute instead? It was the best of diversions.

You never knew when it would come in handy, either, as in the 1979 CTV renewal hearing:

CRTC CHAIRMAN PIERRE CAMU: There is an institute called The Children's Broadcasting [*sic*] Institute.

MURRAY CHERCOVER: We're one of the founding members, sir.

THE CHAIRMAN: And I was going to ask if you continue to subscribe and support that organization?

MURRAY CHERCOVER: Yes, sir.

THE CHAIRMAN: Thank you.

Thank you, indeed.

3

Provincial Possibilities, Federal Paranoia

JUST AS THE CRTC had turned its back on expansion of public television through the cable-subscription mechanism, now, in the mid-1970s, Ottawa paranoia blocked the expansion in another way. For the allocation of revenue from cable was not the only way of underwriting new public television — and genuinely Canadian television — and adding strength and variety to program production. Another sector altogether could be created through provincial broadcasting companies, similar to the CBC in their funding mechanisms and "arm's-length" relationship to the legislature. Over time these provincial broadcasting companies could begin sharing programming and could create a decentralized national network.

The first national network in West Germany, the ARD, had this structure. Broadcasting in West Germany fell under the jurisdiction of the *länder,* equivalent to Canadian provinces. This jurisdiction was jealously guarded, not least to prevent the central government from fashioning the nation's broadcasting after its own single conception of the country. One of the historical reasons for returning jurisdiction to the politically various *länder* after the war was Hitler's capture and perversion of radio in the 1930s.

The programming for the ARD network was supplied by each of the *länder*'s autonomous broadcasting corporations, in co-ordination with the others, and roughly in proportion to their respective populations. The second West German television network, though centralized, was also under the jurisdiction of the *länder.*

In Canada, however, federal order-in-council P.C. 1972–1569 limited provincial broadcasting organizations to educational broadcasting. The Association for Public Broadcasting in British Columbia (APBBC) sent a representative (myself) to Ottawa to present the case for broadening the order-in-council so that, at least, provincial broadcasting corporations could produce and distribute general programming and contribute to the television mainstream.

My arrival was like entering a plague-infested city without a vaccination. The whole of official Ottawa was infected with a virulent case of paranoia about Quebec. Those isolated individuals who were not infected pretended they were, for their own safety. Every man and his bureaucrat fiercely guarded the gates of federal jurisdiction: not an inch to be surrendered anywhere, least of all in communications.

In actual fact, broadening the order-in-council would not have involved jurisdiction. Jurisdiction would still remain with the federal government and the CRTC. This point did not take into account, however, the paranoia of the place. The idea of broadening the order-in-council immediately led to thoughts of a Quebec broadcasting corporation becoming a propaganda tool of the provincial government (remember Duplessis!), agitating the public against federal rule and alienating Quebeckers from the country. Wherever jurisdiction lay in a formal sense, went the argument, a de facto jurisdictional beachhead would be established by such broadcasting. It was best not to risk anything. The hallowed, mythic tradition of a national broadcasting system binding the country together — the tradition behind the founding of the Canadian Radio Broadcasting Commission (CRBC) and the CBC — would be violated, went the argument further. The country would be balkanized (a key word in the Toronto-Ottawa "national" vocabulary). Canada would be broken up piecemeal. Other associated nightmares would ensue. Left unsaid was the assumption that what was good federal policy for Quebec was good for the country. As for western Canadians, they were somehow incomplete Canadians, removed from the heartland, and not altogether to be trusted on their own, either.

In television, Quebec already had two "national" networks of its own, Radio-Canada and TVA. As long as separatism was part of the Quebec scene, it was going to be part of Quebec's broadcasting. Only a witch hunt or censorship could make it otherwise (the federal government did attempt to get a witch hunt in Radio-Canada going, in 1977). The Parti Québécois was duly elected in 1976, despite the little-known order-in-council standing on guard against the balkanizing brainwashing of Quebeckers by a provincial broadcasting corporation.

The definition of "educational television" by a provincial broadcasting corporation also proved to have plenty of stretch to it. OECA (later TV Ontario) began running imported double features Saturday nights. Radio-Quebec, which came on stream later, would run mainline public affairs, hiring away one of Radio-Canada's stars, Pierre Nadeau. In the early 1980s it planned to do news and was only stopped by budget considerations. This allowed plenty of room for propagandizing, if it was going to happen, although a network subordinated to government propaganda would quickly lose its credibility.

The only thing the order-in-council blocked was the leading idea that provincial broadcasting corporations, in addition to an educational service, could and should develop new general programming and ultimately a new national network, producing entertainment, comedy, variety, adventure and drama, and beefing up the anemic Canadian structure. CBS and ABC situation comedy, yes; situation comedy by a British Columbia (and Alberta, and Saskatchewan) broadcasting company, no, in the name of keeping us Canadian. It didn't meet the southern Ontario/Ottawa patriotism test.

The real subversion of any popular cultural sense of Canada among Canadian television viewers, or at least English-speaking Canadians, was not that television in Canada was separatist, or so-called federalist, or potentially separatist, or exhibitionist, disestablishmentarian, square, circular or triangular, but that it was largely American. The preposterousness of putting genuinely Canadian broadcasting organizations in a straitjacket because they were provincial crown corporations — while American programming on "Canadian" stations, quasi-American programming tailored for sale in the U.S., and American channels themselves flooded the country — did not occur to Ottawa.

A young, intellectually regarded policy official of the Department of Communications explained that the order-in-council could not be broadened for a British Columbia broadcasting authority because Vancouver did not have a Toronto *Globe and Mail*. The argument was as follows: The arm's-length separation between the CBC and the federal government stood up, though not without difficulty and incident, because the watchdog *Globe and Mail* was always around to catch and report any infringement. Without a *Globe and Mail*, a similar relationship between a British Columbia broadcasting organization and the provincial government of the day would quickly give way. Therefore, the order-in-council could not be broadened.

Sans le Globe and Mail, *le déluge*, eh?

British Columbians and, by implication, other western Canadians were, it was understood, somewhat unfinished and backward, and could not be trusted to conduct their own affairs properly without the tutelage of southern Ontario civilization and institutions, of which the *Globe and Mail* was a shining jewel. Unfortunately, western Canada, unlike Ottawa, had the misfortune not to be part of southern Ontario. It followed logically that the only adequate federal policy for western Canada was to keep the place in line and stop it from trying anything fancy.

4

The CBC, Part I —
The 1974 Hearing
SOUND AND FURY
SIGNIFYING NOTHING

THE HEARING ON THE CBC in 1974, really a hearing on the CBC's English-language television network, was a great spectacle. Participants felt that history was being made. Juneau and Boyle were at the height of their influence. Juneau, according to *Time* Canada, which was the newsmagazine authority in colonial Canada in those days, "had become a kind of cult hero to broadcasting's dissatisfied programmers." The CRTC was "powerful, innovative." Another journalist remarked admiringly, "In that crowded, restless basement motel [convention] room in Ottawa, it was Juneau who looked most at home, most right for the set, most clearly the man who was doing what he wanted to do in just the way he wanted to do it." Eloquent and intelligent statements were made at the hearing about the nature of broadcasting and about public-broadcasting ideals, impassioned statements as likely as not. The widespread feeling was that the collective conversation was a brilliant slice of cultural and social history, something stirringly Canadian, typical of the best instincts in the country.

There had been the usual criticism through the years, as always, from the ideological enemies of the CBC, woolheads who wanted to get rid of it, who considered the idea of public broadcasting an abnormality and a waste of taxpayers' money. The people who were now raising their voices, however, were the supporters of public broadcasting in Canada and of the idea of the CBC, who were dismayed and angry at what had happened to the English-language television network as they saw it — the loss of its public-broadcasting spirit, the intrusion of commercials, the domination of prime time by American programming, the decline of drama, the absence of investigative boldness in public affairs, the alienation of the artistic community, the ignoring of the Canadian cultural explosion around them, the gathering up of power by administrators at the expense of producers, the loss of intellectual fibre, the blandness and timidity and triviality.

In 1973, the year prior to the hearing, *Saturday Night* showed its preoccupation by publishing four separate and not quite complimentary articles on the CBC, among them "A modest proposal: ban commercials from the CBC," "Incredibly, the CBC gets worse and worse" and "Something has gone very wrong with the CBC." *Time* Canada, giving the hearing the cachet, imprimatur, seal, sign, emblem and stamp of a first-class historical event for a second-class country, did a multipart cover story on the hearing, "What Ails the CBC?" This included a subsection that applauded CBC radio, an interview with award-winning television producer Norman Campbell and a two-page stinger by Mordecai Richler, plus the feature-story *tour d'horizon* and the issue's "Letter from the Publisher," which told how the story was done.

The CBC was not just an ordinary network that generated ordinary expectations. It was part of Canada's soul, "the central nervous system of Canadian nationhood," as Juneau put it. "It's a Holy Grail kind of thing," said Norman Campbell.

Knocking the CBC was a national sport; it was regularly put through the wringer. It was particularly good fodder for politicians. "The most damned, slurred, supported, inquired into, ignored, blamed, upheld, detested, praised organization I know," former CBC president Alphonse Ouimet once put it. "The CBC's roots are being pulled up and looked at every day," remarked Graham Spry unhappily.

There were 305 written submissions to the hearing on the CBC, of which twenty-nine intervenors were chosen to give oral presentations. The proceedings went on for five days. Graham Spry was there as part of a delegation from the Canadian Broadcasting League, the resuscitated successor to the Canadian Radio League. The performers', producers' and technicians' organizations were there in force. Cabinet ministers representing the Maritime provinces appeared. So did Pierre Berton. The most heavily publicized intervention, and the most biting criticism, came from the Toronto-based Committee on Television, which included Robert Fulford, Patrick Watson, Kirwan Cox, Stan Fox, economist Abe Rotstein, filmmaker Allan King and television columnist Morris Wolfe. The four main native organizations in the North sent a representation. The largest delegation was from the Council of Canadian Filmmakers.

There was frequent and outspoken criticism of the CBC's carriage of commercials. The advertising industry defended the commercials. The British Columbia Committee on the CBC, an "ad hoc group . . . united in belief in the principles of public broadcasting," had canvassed newspaper readers in the province through letters to the editor and received 350 replies, which they analyzed and presented to the commission along with their own recommendations for a decommercialized and decentralized CBC.

Juneau's opening remarks to the assembled public were inspiring:

A strong CBC is vital to the health of Canadian broadcasting . . . It is clear that any weakening of the national service, as it is called, would pose a threat to the entire Canadian broadcasting system. Conversely, efforts to revitalize the Canadian broadcasting system . . . cannot succeed without . . . an innovative CBC, sensitive to changing needs, while still constant to original principles of service . . . This hearing is *not* an investigation of the CBC. Let us hope that Canadians have had enough of this peculiar sport. It will be, I hope, a careful and serene discussion of Canadian national broadcasting service, of its mandate, its philosophy, its accomplishments, its future orientations.

The decision subsequent to the hearing was not an ordinary decision, either. It was a 245-page bound book which included a 145-page summary report of the hearing and an honest-to-goodness book title, *Radio Frequencies are Public Property,* taken from the glorious Section 3 of the Broadcasting Act. The report led off with an epigraph from St. Augustine's *City of God:* "A nation is an associaton of reasonable beings united in a peaceful sharing of the things they cherish; therefore, to determine the quality of a nation, you must consider what those things are." The text was sprinkled with apt citations from intervenors and from others including Northrop Frye, T. S. Eliot, John Grierson and Walter Lippman. The first section, "Mass Medium or Public Service," was a wide-ranging philosophical discourse on public broadcasting, containing a critique of one of Juneau's *bêtes noires,* "merchandising strategies," otherwise known as "techno-economic considerations" and the "industrial image manufacturers."

"Commercial activity deflects the CBC from its purpose," the report declared, "and influences its philosophy of programming and scheduling. It must, in the commission's considered opinion, be reduced or even eliminated entirely." And again: "There is no justification, except for the constraints of advertising and the imperatives of mass concepts, for the exaggerated predominance of American entertainment programs on the English-language television service during prime time hours and the relative absence of the best programs from the rest of the world."

While underlining its agreement that the CBC should continue to provide a "popular" broadcasting service with light entertainment and sports in its programming mix, the commission stated that the CBC should get away from mass-programming limitations and strive for cultural originality, although that might mean smaller audiences for many programs (even a small audience for a national television service was huge compared to the concert hall or the theatre). It should maximize "not the audience for every program but the viewers' chances of discovery, understanding, participation and cultural development."

The decision, or rather the book, was an extensive public-broadcasting statement. The chairman was getting a lot of his favourite thinking off his chest. The key conditions of licence renewal for the television network were that commercials could not exceed eight minutes for any clock hour (compared to the previous twelve minutes) and that starting in 1976, commercial time per clock hour would be further reduced by one minute per year. When CBC president Laurent Picard refused to go along, and the issue was referred to Cabinet, the Cabinet found against the CRTC. This conflict of the two French-Canadian "titans" — Juneau and Picard — was widely commented on.

The greater the intensity and excitement and distraction, however, and the more extensive and portentous the media coverage, the greater was the illusion. The event was mostly a wasted exercise. The energy and emotion of the hearing was, magnificently, on the wrong question, and the CRTC would not let it be any other way. Reform of the CBC was not the main public-broadcasting issue; it hadn't been since CBC radio was overwhelmed by the proliferation of commercial licences following World War II. The question that had to be dealt with concerned additional commercial channels and the expansion of the broadcasting structure. With so many commercial channels available and fragmenting the audience, any changes within the CBC's single channels in each language would have only a marginal effect on the overall television picture and on the range of programming possibilities. A perfect CBC would not make much difference. Trying to make the CBC's single channel do everything that a national service should be doing, and everything that the commercial operators refused to do, or could not do, was defying reality, as the BBC knew. A publicly funded national network, moreover, needs to maintain a large audience presence, for its long-range financial and political survival. With only one channel, its programming flexibility is severely limited. It can show some leadership in pioneering new kinds of programming with smaller audiences, and fighting against mass viewing habits, but not much. The CBC in 1974, already hemmed in by competition from a multiplicity of American channels plus a quasi-American commercial network in Canada, had perhaps the least elbow room of any mainstream public television network in the world.

Getting rid of commercials, which should have been done for its own sake, would not have changed this situation. The BBC had no commercials. When its television monopoly was broken with the introduction of a commercial network whose ratings success pinned the BBC against the wall, the BBC fought back to regain its large public presence, for its own survival as a national service, just as surely as if a commercial imperative were forcing it to do so.

The much ballyhooed hearing on the CBC, engineered and conducted by the CRTC on the wrong question, was like fighting a brilliant battle very successfully on the wrong front while the rest of the army elsewhere was being overrun. The whole kit and kaboodle, including the excitable journalists, were furiously, foolishly refighting the public-broadcasting war of 1932 when there was only one front, a war that had long since become obsolete.

The combative soldier, Laurent Picard, was refighting the old war, too, as blindly as the others. "I'm not going to let the CBC be pushed around like [former presidents] have done," Picard vowed before the hearing began. But Picard's manoeuvring on the single-channel front was as passing in importance as Juneau's. The CBC should have been making the argument for a noncommercial third service to take the pressure off its first service for all the specialized and pioneering Canadian programming that its public-broadcasting critics insisted on; in other words, to secure the first network's mainstream position by adding another channel on its flank. Without a CBC-2, or a similar public channel governed separately, the CBC would always be hopelessly trapped, as Picard was in 1974, no matter how stubbornly he made the mainstream argument. As long as the CBC remained on the one-channel defensive, it was cornered. It needed to go on the offensive.

The key to going on the offensive lay in making the general case for noncommercial financing: politically securing the premise that noncommercial funding is a normal and proper way, and also an extremely efficient way, for financing television in general, *all kinds of television*. Financially, the case was incontrovertible; the waste of the commercial-funding mechanism could fairly be described as a scandal. Socially and culturally, the case was also immensely appealing: Noncommercial broadcasting dispensed with commercials. It was as good a case as they come, waiting for somebody with leverage to take hold of it.

Noncommercialism had another, tremendous potential in Canada for a mainstream network. By developing the CBC as a commercial-free, public-interest, broad-gauge consumer and environmental network, with the absence of commercials providing a continuing characteristic cachet, the CBC could have recovered some of the populism that was part of its beginnings. Moreover, this populism was a connection with the public that commercial television could not replicate.

The CBC reverted to defensiveness instead, caving in to the prejudice about public financing. The corporation tied itself into a straitjacket of its own making and kept on pulling the knots tighter. The straitjacket worked like this: If the CBC did not attack the question head-on, the prejudice about public financing would go largely unchallenged. If it went unchallenged, the CBC was vulnerable to the charge that it should not be carrying programming that could be looked after by commercial operations, programs like sports or imported entertainment. It should devote itself to programming that commercial operators would not touch, particularly minority "cultural" programs, and get out of other areas. This, however, threatened to dislodge the CBC from central ground. To prevent being dislodged, the CBC desperately hung on to the commercial financing, or part-financing, of this "mass" programming. Keeping commercials and keeping this kind of programming in the schedule became, in this trap, one and the same thing. But keeping the commercials required higher reasons for doing so, which had to be made up and argued for enthusiastically. George Davidson, Picard's predecessor, was typical. "I wouldn't like to see the corporation entirely removed from advertising," he once said, "because it keeps us in touch with the real world — and off Cloud Nine." There were similar statements made during the Davidson and later the Picard eras. They reinforced the prejudice against public financing. Each time the CBC then applied for an increase in its public appropriation, it had to hurdle the prejudice it had so freely contributed to. In terms of getting everything it needed, it never did, which forced it to rationalize its commercials ever more doggedly, which prevented it in turn from prosecuting the case against commercial financing (if anybody, by chance, happened to be so inclined).

This backing off from the general attack against commercial financing also left commercial television home free against the CBC and its mainstream position. The licensing of new commercial channels was also left wide open. The creation of additional public channels, which an alert CBC mainstream strategy would have made a priority, was, by the same token, thrown into check. You could not argue very well for additional public channels, or even think about them, without taking up the case for public financing across the board. The CBC, in its homemade straitjacket, just huffed and puffed instead, and the CRTC, like a Beckett character stuck in a garbage can, huffed and puffed righteously at it.

5

Global on the Brink

THE UNIVERSE UNFOLDING as it should, the moment that the CRTC officially issued its pontifical treatise on the CBC, March 31, 1974, was also the moment that the original Global Communications was coming apart in financial pieces, many millions of dollars of them, because of a forecasted advertising shortfall of $5 million for the first seven months. Global had been on the air for just three months. The Toronto Dominion Bank, while not calling in its loan to Global, cut off further credit. Global was insolvent. Al Bruner had already been replaced as chief executive officer. "We have a totally unco-operative bank with a lack of understanding of broadcasting," complained his replacement, the chairman of the board. "I think it's terribly unfortunate that the bank would put the gun to Global's head." Nobody bothered to ask why these brilliant financiers and broadcast visionaries, as they were portrayed in the breathless 1972 Toronto newspaper reports, had made the mistake of dealing with an unco-operative bank that did not understand broadcasting. People in the know were watching the frantic rescue attempt as meetings were held day and night with first one and then another prospective buyer. A Stephen Roman-Dennison Mines proposal foundered. A City-TV proposal was tentatively accepted and then rejected.

On March 31 a deal was arrived at with Toronto-based IWC Communications Ltd. and a Winnipeg syndicate headed by Paul Morton of CanWest Broadcasting Ltd.: $12 million to the rescue, read the reports. At the urgent request of Global, a meeting was held at the commission's office on April 2. On April 3 a takeover hearing was scheduled for April 10, seven days' notice instead of the thirty-five days properly required. Interventions to the CRTC were to be filed on April 8, allowing five days instead of the twenty required by the CRTC's rules. (By the time the notice arrived in the mail in some parts of the country, the hearing was already underway.)

There were fervid interventions of support for the takeover and for keeping Global alive. The government of Ontario, through its minister of communications, sent a telegram describing Global as a valuable alternative to U.S. television and an important provider of employment for Ontarians and Canadians. Pierre Berton intervened. So did ACTRA (the Association of Canadian Television and Radio Artists). The independent producers, owed money by Global ("the very existence of many, if not most independent producers is at stake"), outlined their situation.

Global's director of information sent in a personal, and emotional, letter. "These people at Global have been on an emotional roller coaster these past few weeks, living and dying the future of Global, day by day," he poured out. "Rapidly changing events, rumour and speculation have caused personal feelings to course from optimism to despair innumerable times . . . Yet few have abandoned ship."

Henry Comor, the nationalist broadcaster and producer, wrote a similarly impassioned letter, quoting the Hungarian composer and teacher Zoltán Kodály: "Unless the modern man can develop simultaneously a strong and healthy relationship with his own family, his own people, his own background, his own language, music and dance, and cultivate the abstractions and idioms of his own age, he will never be a balanced human being but will remain forever dazed and confused, a prey to passion and prejudice in the face of the jungle onslaught of modern life."

"I have the right to demand that Global continue to exist — in its present format — for myself, my family, and my community," finished one long, telegrammed encomium.

"Global is needed to stop ever increasing Americanization," said another.

"It would be a blot on Canadian broadcasting to thwart the progress after three short months," said another.

"At last a television network is making me feel Canadian," said another.

The Braden Beat, a consumer program, broadcast a special program reflecting the concern of some viewers about Global's future, and 5500 letters and telegrams came in. A petition organized by a group of London, Ontario, students was delivered to Toronto by hand to get there in time; the students had borrowed a teacher's car and his oil company credit card to make it. Some people sent money, $2 bills, $5 bills, $10 bills and cheques.

It was revealed that Global's ratings were improving rapidly. Explanations were offered claiming that the difficulties were short-term in origin. Allan Slaight, head of the IWC syndicate, was positive and optimistic. The existing corporate vehicle, IWC Communications, owned three cable systems as well as three radio stations.

"We know these to be good people of high integrity and we believe their intentions are most honourable," said the chairman of the board of Global Communications of the IWC syndicate at the emergency takeover hearing, held in Ottawa.

"We come before you today in an effort to rescue Global and still maintain the high principles and ideals and, in the main, the commitments of its original founders," responded Slaight.

If there was one thing Slaight "agree[d] with wholeheartedly" and was "naturally committed to maintaining," as repeated mentions at the hearing underscored, it was "Global's original premise and the basis on which the Global licence was issued."

"We, and I hope you appreciate our sincerity in this area, have strongly underlined that we believe in and will go forward with it, the Global program commitment . . . Our intention to conform and to improve on the original Global mandate is sincerely presented," said Slaight.

"Do you think," asked Juneau at one point, "that you will still be able to boast about your contribution to performers and independent producers a year from now?"

"If the network is on the air, yes," said Slaight.

Seymour Epstein, an engineer who had helped to shape the take-over deal and who became a minor partner in the takeover syndicate, offered that "Global can be one of the most significant developments in broadcasting in many years. It can become a truly national network." Epstein was the former director of the planning and development branch of the CRTC, and had reportedly worked on the third-service idea while at the commission. Later, on leaving the commission, he showed up as an engineering consultant for Global.

He now described a meeting with spokesmen for the independent producers, one of whose conditions of support was that creditors would be treated honourably. This commitment and others (to retain the "Global concept," for example) were easily obtained from the syndicate, although it was explained to the producers that through the critical period there would likely be a cutback in the number of independent shows. "We want the support of [the independent producers]; we intend to pay them what Global owes them," said Slaight.

Paul Morton, speaking for the Winnipeg partner, later to be incorporated as Global Ventures Western Ltd., spoke pregnantly about how westerners, through this participation, would have "a meaningful influence in what ultimately may become a national network," although how that influence would make any programming difference, where the mutual object was to make money, he did not say.

This was not, however, really a licence hearing. It was a hearing about how to save everybody's skin, including the CRTC's. The spectre of bankruptcy, receivership, stripped shareholders, ruined production companies, desolate debenture holders, belligerent unpaid suppliers and an inconvenienced Toronto Dominion Bank hovered everywhere in the air of the Commonwealth Room of Ottawa's Holiday Inn Hotel. For unhappy debenture holders trying to salvage whatever they could, the trick was to make the CRTC feel as guilty as possible without actually being so gauche as to say the CRTC was culpable. One debenture holder's lawyer was particularly agile in his imputations. "Rightly or wrongly," he said, "the public assumed that the commission and the staff had passed upon the economic viability of the operation. ... The public was led to believe that in some perhaps completely unjustified way ... they had some assurances." Juneau, while denying legal responsibility or the responsibility for individual investment decisions, did volunteer, in all this unhappiness, that the CRTC had a "general responsibility" for the situation.

The truth was that the CRTC did not have any such responsibility. If some promoters wanted to try flying a doubtful scheme and could get financial backers for it, good luck to them, but the CRTC could not be held responsible if the scheme didn't fly. The promoters had coaxed, winkled, wheedled, inveigled and squeezed the licence out of the CRTC in the first place. One only had to remember the remark of one of the commissioners when the licence was first awarded: "You figure out how they can do it, brother." But because of the CRTC's paternalism generally, keeping licensees from going bankrupt was its doctrine, and its own status was on the line. People wondered out loud how the CRTC could have licensed a network that was not economically viable. The commission was supposed to be a lot smarter than that. The commission had already lost face and would lose a lot more if it announced, after the Bruner exercise and so many million dollars down the hatch, that Global was a foolish idea after all. "If Global should fail now," wrote Jack Miller in the Toronto *Star* a month later but with the same train of thought, "the CRTC's financial judgment would fall into disrepute, no matter whose fault the Global troubles were."

The CRTC and its commissioners were in a conflict of interest, making a decision that involved saving their own reputations. If they decided, too, to let the licence lapse despite the takeover offers, the fallout from the financial victims and unkind observers would end up on the CRTC's head. The commission was supposed to be responsible to the Broadcasting Act, not to creditors whose business losses were their own responsibility. Saddling the whole country for the indefinite future with a misbegotten sector of television to satisfy some creditors was not what the CRTC was established for.

"Even if the programming is weakened considerably," Juneau told a journalist while the manoeuvring was going on, "the commission has to consider whether it is much better to keep the network on the air, in part to give a better deal to the creditors." It was duly found that it was much better. The decision approving the takeover was issued the day after the hearing, to forestall further financial disintegration which could have been fatal to the patient. At the insistence of the takeover syndicate, the restriction of eight minutes of commercials per hour, one of the reasons for originally allowing the licence, was loosened to ten minutes of commercials per hour.

Right on schedule, in August 1974, the biennial hosanna to the CRTC and its chairman appeared in *Saturday Night*, succeeding Batten's article in 1972 and Fulford's in 1970. It was another chapter in the heroic saga of cultural patriots and their staff of "able turks" in Ottawa challenging the CBC and withstanding the Global affair, overcoming all roadblocks, manmade and self made. It was written this time by Robert Lewis, who had done the research on *Time* Canada's CBC story. "Pierre Juneau's spectacular leap of faith" was the article's title, referring to faith in Canadian creativity. The magazine had once again missed the point.

In the little-known but illuminating academic study, *Canadian Television Broadcasting Structure, Performance and Regulation*, done for the Economic Council of Canada and published in 1979, communications economist Robert Babe explored the financial background of the Global takeover. This, and additional details from the record, make an interesting picture.

The new Global regime was not long in taking care of the creditors for whom the CRTC had "in part" kept Global alive, although not in the way the creditors or the CRTC had anticipated. One of the first things IWC did was arrange a management contract with Global whereby IWC would provide a chief executive officer and a chief financial officer for $175,000 annually. Shortly after, the new management informed the trade creditors that it had discovered an extra $1 million in liabilities that it was not aware of at the time of the hearing, and was thereby forced to renege on its obligations to them. Otherwise, it declared, it was immediate bankruptcy or receivership. In case of receivership, the bank loan and IWC's commitment to that point would be secured by the assets. The unsecured trade creditors, however, would be at the end of the line. They included the independent producers, who had so understandingly and thankfully supported the takeover and whose development and prosperity had been one of the major reasons for originally allocating the licence.

IWC offered the creditors either 25 cents on the dollar or payment that did not begin until five years later and did not end until nine years later, with no interest in the meantime. The cash offer was a deal the creditors, hemmed into a corner, could hardly refuse. Global's annual report for the following year showed a "gain on settlement claims of unsecured creditors" of a nice $2.7 million.

Izzy Asper, who became through CanWest one of the major shareholders of Global, and was a financial analyst, wrote in a memorandum at the time of the takeover that "Global was given an historic licence valued by many people in the industry at around $15 million." The potential profits were considerable. The risk to the IWC-Ventures syndicate, on the other hand, in taking over this prize was, according to Asper, "almost nominal." He described the takeover as "a very outstanding opportunity."

The syndicate's projected capital injection consisted almost entirely of capital debentures bearing an interest rate of 1.2 per cent above prime, which would cover the rate the syndicate itself would pay for the funds. This investment of $8.9 million, unlike the case if it had been equity, was secured against Global's assets, subject only to the prior claim of the Toronto Dominion Bank.

Although the syndicate's investment was in secured debentures rather than in equity, the agreement nevertheless also gave the syndicate a hammer lock on profits: a sum equivalent to 30 per cent of *pretax* profits in 1975, declining by one point a year to 20 per cent in 1985 and thereafter to 1998. Payment for this would come out of *posttax* profits, or from 60 to 40 per cent of the posttax return for the respective years, given a full rate of taxation. In addition, the

IWC-Ventures syndicate had a ten-year option to purchase most of old equity, beginning at bargain-basement prices, and capture the bulk of the remaining profits as it saw fit.

By contrast, the original debenture holders who had provided $10 million lost the security of Global's assets against their investment, as the debentures of the syndicate would have prior claim. The debenture holders also had to make other concessions, like deferring three years of interest. They had to give up most of their rights to convert their preferred shares (five shares to a unit) to common stock, losing possible future gains. The conversion rights had been an important selling point in the issue of the low-interest (6 per cent) debentures. Simultaneously, the debenture holders' voting power was reduced by four-fifths, from a majority interest, taken together, to a minor one.

"Treatment of trade creditors and debenture holders," wrote Robert Bube dryly, "appears to have been harsh in light of Asper's judgement, at the time of the takeover, that the licence was worth $15 million and that 'the downside risk is almost nominal.'"

At the same time, the equity holders who had negotiated the deal — CanPlex (the company headed by Al Bruner), Maclean-Hunter and Odeon Theatres — would recover cash when the new syndicate exercised its share-purchase options and bought their stock. If all the options were eventually taken up, CanPlex, it appeared, would get back virtually the whole of its investment for the 94 per cent of their shares that were involved. The other two partners would recover approximately half their outlay.

There had been another offer for Global, from City-TV, possibly more advantageous to the debenture holders, with the old management block reduced to a small shareholding. This was appropriate to the circumstances. Normally in the case of a company on the verge of bankruptcy and with its stock plummeting in value, it is the equity holders who first take the loss. After the takeover hearing a third proposal was made.

The debenture holders, however, never got to choose among the alternatives. The CRTC would not entertain a competing application, from City-TV or any other party, for the public licence. Without the right to have an application heard, everything else was irrelevant. The commission would entertain an application only from the party chosen by the outgoing management group. The outgoing management group chose the deal most favourable to itself. The debenture holders were hamstrung. The IWC-Ventures syndicate and Global's equity shareholders were home free.

The commission's stated policy was to not "intervene in normal bargaining between the current holders of the assets or shares of licensees and would-be purchasers." This blithely ignored the fact that the bargaining involved the public licence, under the CRTC's supervision, as well as the privately owned assets. By refusing to hear alternative proposals, the commission let the outgoing management use the public licence, which it did not own and which was meant for public service, to its own private financial advantage. Most important, the commission's action meant that the public did not necessarily have the best new licensee, although the best possible broadcasting was supposed to be what the CRTC was about and what licences were for.

Having manhandled the bondholders and the creditors, the new Global management then went on to hack away at its Canadian programming, dropping "every show which was designed to justify its existence," as Toronto *Sun* columnist Bob Blackburn rasped. "It's appalling that this situation should have come to pass and that Global, which promised so much and is now delivering so little, should be getting away with it."

The bitterness from performers and writers flowed freely. The Toronto *Star* carried a report of a tearful Barbara Hamilton, whose show, *Shh! It's the News*, with Don Harron and Catherine McKinnon, had been unceremoniously dumped, after they had "worked for peanuts" and put up with impossible conditions for two years to keep the show alive. "A sad day for Canadian performers and for the naive hopes that accompanied Global," intoned Blaik Kirby in the *Globe and Mail*. "What, if anything, will be done about it by the Canadian Radio and Television Commission?" grumbled Dennis Braithwaite in the Toronto *Star*.

News time was expanded to twice the hours Global was doing in its original schedule, in order to help fill up the prime-time Canadian-content requirement. This was the opposite to increasing the proportion of Canadian programming in drama and variety, and avoiding duplication of programming provided by others, for which Global had been created. ("No, we will not be into drama," Slaight said, "unless there is a guarantee up front of a U.S. or an international sale.")

Global entered the bidding against the CBC and CTV for other American programs to run in prime time, which brought it greater revenues, which were applied to the bidding for American programs, helping to raise the cost of American programs in Canada and forcing more Canadian money out of the system and away from Canadian producers, into the United States.

"I said a few things [at the takeover hearing] that have not proved to be correct," said Allan Slaight modestly.

Amid all the crying and groaning of dismayed nationalists, displaced performers and grumbling journalists, everybody overlooked the fact that the original Global schedule, whose passing was being mourned with so much distress, was thoroughly American despite the bold attempt at some Canadian program production. Half the evening schedule, for a start, was non-Canadian, almost all of it American. Most of the American programming was in the peak midevening hours, with the news and the variety talk show *Everything Goes* on the edges, just like CTV Canadian content. The American shows were spotted head to head with their appearance on U.S. channels to take advantage of cable simulcasting.

But *Everything Goes*, an hour and a half every night, making up almost half of the prime-time "Canadian content," was itself largely American. Its main host, comedian Norm Crosby, was American. Most of the guests were American. The show was modelled on an American program — *The Mike Douglas Show*. As Morris Wolfe observed on the opening of the network, much of the talk on the show assumed its audience was American, and when expatriates were introduced, they came across as "Canadians who've made it" . . . to the big time in the States.

"Partly because of the belief that Canadians can't do variety, partly because Global is eager for U.S. sales," wrote Wolfe, "there are so many Americans on *Everything Goes* that someone stumbling across it while flipping the dial would suspect the programme had been produced in the States." Structurally, Global was a ringer from the beginning.

In the fall of 1975, after all the Global babies had been massacred, the CRTC held a hearing on a minor matter of Global transmission in the Windsor area and used it to review what had happened. Harry Boyle was now acting chairman. The conventions of the ritual were as follows: Slaight had to tell the commission that the economics of even a watered-down "Global concept" were prohibitive, which he did pointblank. "We cannot both meet the mathematical requirement and also mount and stage and pay for the kind of programs that were in Global's original schedule," explained Slaight. The ritual called for him to beg the commission for relief and to outline the cataclysmic effects if relief were not provided, such as broken investors and employees out on the street. At the same time, he had to reassure the commission that it had not made a mistake in licensing Global, although if the original concept had reflected reality as described by Slaight, the licence would not have been issued. Logically, this was an impossible job. However, if everybody, including the licensee, agreed that the licence was a mistake, what more could be said? They would all have to go home.

Even a thinly disguised, qualified defence by the Global syndicate of the licence's existence, on the other hand, allowed the commission to indulge in scepticism and toughness — to let off steam and generally play the critical body. This was the only useful function of the hearing: to allow the commission to bang its head against the wall, after which it might feel a bit better.

D. McDougall [part-time commissioner]: In the cold light of today, do you think it might have been a mistake in the first place for the CRTC to licence Global?

A. Slaight [ritualistically]: No, I think Global definitely has a place and I think we're proving that.

H. Boyle [chairman]: To follow up, and I think, you know, it's a nasty question, it's nasty for us as well as for you: Are you saying in effect that there is really no hope for the Canadian broadcasting system if it's to follow the dictates of the Act that says we have to have a predominantly Canadian broadcasting system?

A. Slaight: No, I just think that there should be three tiers [of Canadian-content requirements, with Global's schedule not being predominantly Canadian].

H. Boyle: You know, we can't go on . . . — I'm sorry to take this out on you, but we have a great many problems in terms of whether or not we're going to have a broadcasting system in this country at all.

A. Slaight: Sure we are.

H. Boyle: And the fact is that this total dependence on the American product really isn't a good enough answer. There has to be some alternative found or else we just have to close up the ruddy shop and let the Americans establish affiliates in Canada.

Of course the next year, 1976, the licence was renewed on expiry. Just to be on the safe side Global, prior to the renewal hearing, introduced a few token independent productions, which even the CRTC did not take seriously. One of the commissioners at this 1976 renewal hearing, Roy Faibish, went to great lengths to put Global on notice for the following hearing, just as the 1975 decision had gone to great lengths to put Global on notice for the 1976 hearing. The restriction on commercials was raised to twelve minutes per hour.

The Association for Public Broadcasting in British Columbia (APBBC), which had systematically put two increasingly detailed financing alternatives to the commission through late 1973 to 1975 — in submissions for Montreal, Ottawa (the hearing on the CBC), Winnipeg, Calgary, Vancouver, Ottawa (the hearing on cable policy) and in a separate correspondence — watched this CRTC moaning and tears with growing cynicism. The commission had not wanted to hear the alternatives, much less take them up.

In the spring of 1975, just a month before the hearing on cable policy, chairman Juneau had delivered a long, gloomy address to the Canadian Association of Broadcasters on, among other things, how cable had failed him. "More than an hour and a quarter of public agony," Blaik Kirby described it in the *Globe and Mail.* It was one of the last of the heartfelt, imploring and altogether useless speeches that Juneau made as chairman of the commission. Boyle was a worthy successor in the tradition. The scope for anguished declarations about the need for an alternative, while ducking an alternative, was unlimited.

Here is another bit of ducking, from the 1975 hearing into Global (Allan Slaight again playing the straight man to Harry Boyle):

H. BOYLE: I have a suspicion that if we have to introduce into Canadian broadcasting a federal subsidy plan for what is now considered commerial broadcasting, that we will have destroyed some of the credibility of our system.
A. SLAIGHT: I agree with that.

Whose system was Boyle talking about? If there had been a parliamentary subsidy for Canadian program production, or better still, public funding of noncommercial programming in specific hours inside the schedules of commercial television — children's programming, for example — we would have added credibility to a system that had very little.

The logical follow-through from the evidence that confronted Boyle and that he admitted and groaned about was to get rid of Global and replace it *in toto* with a publicly funded network. This would not have destroyed the credibility of anybody's system — the broadcasting system or the free-enterprise capitalist system of A. Slaight and, it appeared, of H. Boyle — except the system of privately mining publicly owned broadcasting licences. Great Britain had two general-programming public channels. The Netherlands had two public channels. West Germany had two public channels.

France had three public channels. Sweden had two public channels. Canada's position next to the United States with its commercial channels and its economies of scale provided more of a rationale for new publicly funded television.

Two and a half years after the financial and political legerdemain of the Global takeover, the syndicate having caught its corporate breath, the IWC half of the partnership (Slaight and company) activated a buy-sell clause in the contract for the other half. The partner, Global Ventures Western (controlled by Paul Morton and Izzy Asper, who also controlled CKND Winnipeg), opted to buy rather than sell and took over the company.

IWC's investment in Global to that point was $3.5 million. The deal, approved by the CRTC in the wings, brought IWC $6.8 million, for a capital gain of $3.3 million, or 95 per cent, in exchange for an admittedly negligible financial risk. This was much more pleasant than the 75 per cent loss with which the syndicate had stuck the independent producers and the trade creditors on accounts payable; they were plumb out of luck. What in effect was being sold for this $3.3-million difference was the public licence, whose financial value had been inflated by all the programming games at variance with the licence's reasons for existence. This value was then appropriated by IWC. In this way, money went out of the broadcasting system, and ultimately out of programming, to IWC. If some ambitious citizen converted public property to his own use to the tune of $3.3 million, he would be behind bars. Yet the decision of the CRTC in effect allowed IWC to do the same thing legally.

In 1974, when Paul Morton had invested in Global while at the same time applying for the Winnipeg licence, he had declared himself the patriotic Manitoban, putting the Winnipeg licence first. Morton now moved to Toronto to become Global president. He came from the movie-theatre business and had recently been president of the Motion Picture Theatre Association of Canada, representing the ultimate in foreign-dominated distribution and exhibition systems, and one notorious for its almost nonexistent domestic content. Morton was just the man, in the upside-down CRTC scheme of things, to end up with the Global licence originally allocated for great Canadian pioneering purposes.

6

Cable Rates

THEN THERE WAS THE CASE of the CRTC's magic regulation of cable
rates without any real regulation. The commission, by dint of creative rationalization and intellectual craziness, had managed to convince itself that it was business as it should be done.

One effective way of doing this was to withhold from intervenors
the financial information they needed in order to intervene.

The issue of disclosure was first brought up by the City of Victoria in 1972, during a hearing on a rate-increase application by
Victoria Cablevision. Victoria's mayor at the time was Peter Pollen,
an iconoclastic automobile dealer with a degree in political science
and a principled sense of the public interest. Counsel for the city
asked for an adjournment of the hearing on the grounds that:

> This application is based solely upon a financial requirement and the
> only information that we do not have is financial . . . ; upon requests of
> both the commission and of the applicant, no financial material has been
> [made] available. It is, therefore, impossible to properly prepare the
> intervention that is required by your rules because there is literally no
> material that is available from the portion of the private company which
> is the applicant.

The fact that the argument had to be made gave the encounter
an air of other-worldly absurdity. The counsel also asked for the
right of cross-examination:

> From what I have learned, [it] isn't a habit or custom of your commis-
> sion to permit cross-examination. But may I also submit that when you
> reach the point where the total material that is to be provided must

come from the applicant, how else can it be examined on both sides to the advantage of the commission, unless there is cross-examination with respect to it?

Juneau ruled against adjournment, since the commission, he declared, had "applied its procedures fairly as they are at the moment." Whether the procedures that were applied fairly were fair procedures was not something to be allowed to inconvenience the commission. Faced by a legal counsel for a good-sized city making an obvious point on a ludicrous situation, however, Juneau played the obliging public servant. He announced that the issue had been "raised very strongly" and the commission "should consider this point very seriously" and therefore "it will take this point under advisement and, as soon as possible in the future, make a public statement on this matter."

Juneau subsequently wrote to the Canadian Cable Television Association (CCTA) soliciting their opinion. The CCTA replied that in general cable companies could be expected "to reasonably document financial arguments that the cable company may use to justify basic service residential rate increases." What if the CCTA had written the opposite? The CRTC's job was to regulate them, not to kowtow to them. (The CCTA, in its response, tied its position to a recommendation that increases up to 4 per cent annually would be allowed without a hearing. Since cable companies could build up huge rates of return by adding subscribers and increasing penetration along trunk and branch lines already built, without benefit of a rate increase, an annual 4 per cent rate increase would have been, for many, cream added to cream. They could still apply for more if so inclined.)

Two years went by. The commission sat on its hands. The CCTA then reversed field. In a hearing in Ottawa in 1975 it argued against public disclosure, noting among its reasons that: (1) The CRTC hearing was "not intended by the legislature to be an adversary proceeding" between two opposing parties, with the CRTC adjudicating; (2) There was a danger such information would be misconstrued by those not familiar with the company and its operations; (3) It "would tend to focus public attention purely on the financial aspects of the company's operations"; (4) "Quite often these things are fought not in the commission, they are fought in the newspapers . . . it is a great opportunity for somebody to start a public debate that is going to put the company under tremendous pressure to respond," and (5) "The important thing in the public debate is that it be a controlled debate; I think you would agree and we are all grateful that the CRTC controls the public debate."

No doubt the cable lobby was grateful. Juneau, at the hearing, had suggested to them that an open debate would clear the air for cable companies, and that the current debate was "a kind of stupid public debate" because little data was available to the public. What makes the lobby's response interesting is how it played up to the commission's paternalism, how it indulged the commission with the idea that father knows best. The cable lobby knew the commission's psychology better than most.

In connection with a rate-increase application by London Cable TV, also in 1975, the Consumers' Association of Canada (CAC), represented by legal counsel Andrew Roman, took up the cudgel. Roman wondered out loud about the futility of intervening when the necessary financial information was not available. The commission again deferred ruling on the matter and again proceeded with the hearing. It also denied a motion for cross-examination.

Faced with the threat of legal action by the CAC, however, the commission, in its eventual decision, ruled that the most recent financial statements of applicants be made available to intervenors in future. Three years had elapsed since the motion by the City of Victoria. The commission reserved to itself, nevertheless, the right to make exceptions upon the request of an applicant "when it considers the public interest will best be served by so doing." It declined to make disclosure of financial projections a requirement. And it allowed the London rate increase to go through under the old rules — slipping one more case by, even though the CAC motion had preceded the hearing.

So the Consumers' Association of Canada (CAC) took the CRTC to court, in the London Cable case, the taxpayer footing the bill for the commission's little game. The Federal Court of Canada found that "there was not made available to the [intervenors] as members of the public a reasonable opportunity to know what was involved in the application" and threw the rate-increase decision out. The commission was obliged to hold another hearing on the application, this time with the financial statements in the public file. (Harry Boyle was now acting chairman.) These statements, forced into the open, revealed that in 1975 the licensee had an extraordinary rate of return for a regulated monopoly of approximately 52 per cent before taxes and 25 per cent after taxes. This was well over twice the rate of return allowed Bell Canada. For the 1975 financial year, further, London Cable had paid out in dividends almost half again what it had earned, so it did not need a rate increase to solve any short-term cash problems, either. In any case, long-term borrowing or the issuing of new shares, rather than an exaggerated rate of return, is the proper method for a regulated monopoly to acquire cash for unusually high capital expenditures whenever its own reserves are insufficient.

The commission allowed the previously awarded rate increase anyway. Nothing in the court decision obliged it to do otherwise.

Behind this strange unregulation by the regulatory agency was a fanciful paternalistic strategy indicated in various commission documents. The strategy arose out of a general preoccupation with cable. Juneau and other cultural nationalists were tormented by cable, whose hardware power was confounding the attempt to defend Canadian objectives. "Technological cancer" was what Juneau had called it. His speeches and commission documents returned again and again to this theme of how the cancer was "enlarging the pipes" and "broadening the receiving funnels" without a matching increase in our capacity to produce. Cable was the most disturbing part of this "natural tendency of hardware, tools, and machines to proliferate as a result of technological and marketing pressures."

The technological cable creature had already humbled the CRTC in 1970, when pressure forced the commission to allow microwave relays of distant American signals to cable systems in the interior of the country. While making this retreat, the CRTC joined to it a counter-strategy. One of its original components was to reject a rigid control of rates. Any such control would keep rates down, which would result in the maximum number of people subscribing to cable, not what the commission wanted to see happen. A parallel idea was that cable should be financially strong in order to pump money into Canadian stations whose audience was being fragmented by cable, and thereby build up their Canadian programming. This idea of cable contributing to Canadian program production was floated by the commission at the time. The CRTC also envisaged the possibility of the "over-the-air" system being so strengthened by cable payments that it would become a better substitute for cable, forcing cable rates to decline automatically by the pressure of "market conditions."

With the same kind of thinking, the commission drew up a formal list of fourteen criteria for judging rate-increase applications, in September 1974. The largest group of criteria had to do with quality of service and meeting CRTC policy requirements. "The role of

the licensee in the development and strengthening of the broadcasting system in its area" was one of them. "Economic need" was
another general category. Intervening against a rate-increase application with these criteria was like walking on quicksand. The public
disclosure of financial statements, finally forced out of the commission, hardly helped at all. If a licensee met policy requirements
which it was supposed to meet anyway, for example, did it get an
extra reward, and by how much, and how was the measurement
taken, and above all why? Nobody ever said. The criteria did not
tell. Rate-increase hearings, which should have been simple exercises in financial measurement and analysis, became drawn-out,
loose-ended perambulations followed by secret calculations that
only God and the CRTC knew, if indeed anything resembling a calcu-
lation came into it.

The cable industry soon picked up on this procedure. Applicants
would systematically go through the list of criteria, throwing in
anything they could think of in the categories they wanted to use.
One of the favourite listed criteria in later years, as licensees discovered its bureaucratic possibilities, was the "introduction or improvement of community programming." This allowed the applicant to
fill his application with program schedules, lists and descriptions of
programs, letters from individuals and organizations involved in
community programming, and assorted solicited opinions. This
information had nothing to do with rates, but it gave rate-increase
applications the desired weight and thickness, purporting to indicate by bulk and detail that there was more to rates than getting a
fair rate. Chitchat about the community channel could also be
expanded during the hearing, sometimes fortified with a videotape
presentation. Commissioners who found a balance sheet too complex could nevertheless show their conscientiousness by asking
questions about the community channel until heads began to nod.
Any intelligent stranger observing this exercise would have
assumed he had run into a renewal hearing and wondered at what
other hotel or motel convention room the scheduled rate-increase
hearing was taking place.

The encounter between the Consumers' Association of Canada
and the CRTC over the rate-increase application of London Cable
TV was as good an illustration of this phenomenon as any. The CAC,
in a careful, iron-clad argument, demonstrated that London Cable,
as a regulated monopoly, was already overcharging the public. London Cable freely admitted that it did not need any more money. It
just said that it deserved more as a reward for its efficient

"entrepreneurship," as if it were not a regulated monopoly. It pointed to various other "criteria" in the CRTC's list, which it had met. The CRTC, in order to defend the validity of its criteria, if nothing else, was in return obliged to give at least something of a rate increase to the applicant. Incredibly to the CAC and to common sense, that is what the commission did, making London Cable's extravagant rate of return still more extravagant.

The 1974 list of criteria functioned as a bureaucratic device for gifting cable company shareholders at the public's expense — quite appropriate to the scheme of things, too, for gift-giving is a typical element of paternalism.

An intervenor could not get much more of a grip on a licensee who argued "economic need," either. The only way of calculating what a cable company needs financially is to establish what "rate of return" on investment cable operations of that category require to attract investment capital and to support borrowing (together, the "cost of capital") in optimum balance. Anything above that is not needed. The CRTC's criteria, however, avoided any mention of rate of return. Unlike the CRTC's regulation of Bell Canada and B.C. Telephone, when it took over that function in the mid-1970s, "rate-of-return" regulation was never allowed to sully the face of the cable industry. "Economic need" was thereby a phrase that did not necessarily mean economic need, or anything in particular other than that a licensee was asking for more money and was using the phrase. But it sure sounded significant.

Licensees had a field day. The rate-increase applicant would go through standard diversionary arguments like rising costs due to inflation, or the burden of new expenditures — arguments as predictable as the forms of kabuki theatre — to enhance the fiction that rates were too low. This would be done with all zest and resplendent detail even if the rate of return was outrageous — double or triple what other regulated monopolies were getting. An intervenor making a rate-of-return analysis, if by happenstance such an intervenor appeared, would put it forward blindly into the CRTC murk from which it would never again be seen or heard.

One of the fictions for avoiding rate-of-return regulation was that cable was not a public utility, although in some cases penetration was in the 90 per cent range. This was part of the cable industry's rhetoric of being a risk-taking, pioneering enterprise for which no reward was too great. It also happened to cater to the CRTC's obsessions. Juneau strongly opposed the very idea of cable as a public utility because it meant for him that everybody would consider themselves entitled to cable. This would call for, in turn, ever more costly additions to capital plant in order to bring cable, and American television influence, into the remotest corner of the country. It would also, according to the Juneau scenario, involve cross-subsidization. Meanwhile, the CRTC's objective of allocating funds from cable into program production would be squeezed out.

The fact that in its technical, financial and monopoly characteristics cable resembled utilities like nothing else did not dislodge this *idée fixe*. Nor did the fact that rate-of-return regulation was called for whether or not cable went under the name of a "utility." It was a monopoly using public licences and under the jurisdiction of a regulatory agency. There was also virtually no risk in cable. Whatever the risk, moreover, it would be reflected in the allowable rate of return, if one were established.

Rigorous rate-of-return regulation also did not interfere with the CRTC's cultural objectives. Rate of return is based on net revenue *after expenses such as the allocation of funds to programmers.* A licensee who allocated funds to program production could ask for a rate increase to bring his return on investment to the appropriate level and be just as far ahead as a licensee who did not allocate funds to program production and had his return limited to the same level.

The commission's strange scheme for the unregulation of cable's rate of return had another flaw. The damned thing didn't work. The trouble was that the commission did not have the jurisdiction to force cable operators to hand over their surplus profits to broadcasters. Cable operators pocketed the extra money instead and complained bitterly about how the CRTC was suffocating them with regulation.

An independently minded regulatory agency would have reallocated cable licences when they expired to applicants who were prepared to distribute surplus earning capacity to program production — and best of all, as proposed by the Association for Public Broadcasting in British Columbia (APBBC), to a new noncommercial programming sector. The CRTC tried instead, under Juneau, to inveigle, coax, manipulate, shame, intimidate and otherwise induce the cable operators into doing what it wanted them to do. But none of the cable operators broke rank and obliged the commission. They just kept pocketing the extra loot and kept on complaining about CRTC regulation, while indulging the CRTC's paternalism on other matters, like the community channel, to keep a good thing going.

Time passed. Legislation obliging cable to do what the CRTC had been unable to inveigle it into doing was not forthcoming. But the mysterious rate-increase ritual by which the CRTC gifted or, on occasion, declined to gift a supplicant remained. Without a rate-of-return rule, contested rate hearings continued to resemble shouting in a swamp.

In the fall of 1975, in the middle of these events, Pierre Juneau made an unplanned exit from the broadcasting scene.

Prime Minister Trudeau's old friend Gérard Pelletier, the minister of communications, was named by Trudeau the new Canadian ambassador to France. The new minister of communications, named at the same time, was Prime Minister Trudeau's old friend Juneau. Although the Liberals never showed any compunction about erasing the line between the civil service and the party, the appointment gave journalists an extra subject to write about.

"Intense, driven, bold, a man with a mission equipped with an uncompromising, crystalline intellect," wrote Richard Gwynn, "also an elitist, an autocrat and a ruthless byzantine operator in the political backrooms ... one of the most intriguing and commanding personalities on the national stage."

"A man whose intellectual, spiritual and political characteristics designate him to replace [Pelletier], one of the three 'wise men' of 1965," wrote Michel Roy in *Le Devoir*.

"The tough-minded antagonist," wrote Canadian Press.

Strangely, it seemed, in the light of the tough-minded image of Juneau, the private-broadcasting spokesmen welcomed the appointment, at least for official consumption.

"I am very pleased with the appointment," said Pierre Camu, president of the Canadian Association of Broadcasters. "He has great experience in the field. It isn't like a new minister coming in with no knowledge."

CTV president Murray Chercover called the appointment appropriate and desirable. The new minister was "uniquely aware of the delicate nature of the total communications industry."

"I hope we will not be losing him entirely," said Bill Stewart, the program manager for the Global Television Network, "but we're definitely losing a brilliant man."

Although in the Cabinet by virtue of the appointment, Juneau had to win the by-election in Hochelaga, the seat vacated by Pelletier. Juneau lost to the Conservative candidate in a major upset. Ultimately, the Conservative who beat Juneau crossed the floor of the Commons and joined the Liberals.

7

City-TV

THE LITTLE STATION
THAT DIDN'T

IF AL BRUNER WAS TOUTED as "television's superman," Moses Znaimer and City-TV were its Cinderella, far from a disadvantage. The capacity for hokum was actually greater.

City-TV, or Channel SeventyNine, was licensed seven months before Global in 1971 as a local independent station that was going to bring a revolution to Canadian television. "The first secret," Znaimer announced, "is to think small." City-TV, the group's application boasted, was not going to be another VHF (Very High Frequency) station like the others, the "established hogs of the electronic frequencies," with their "near uniformity fare" and "in any case, contrary to the public interest." City was applying for a UHF (Ultra High Frequency) licence, whose smaller broadcasting contour would closely approximate the boundaries of Greater Toronto. It was going to be an "[alternative] to mass audiences and mass costs" and would do so by skipping the massive studios and fixed equipment of the "dinosaurs." Other than one "modest" studio, small mobile units would be used: "Channel SeventyNine programming philosophy," said the application, "is based primarily on mobile and remote program origination . . . the studio is . . . secondary." Similar patter was woven through the application. "The Channel SeventyNine concept is based on engineering being subservient to programming ideals and philosophy and to the necessarily constrained budgets of the operation."

The low-powered UHF transmitter required minimal installation space, cooling and electrical power, making it possible to use a lightweight antenna and place the complete installation on the roof of a high-rise. Dispensing with the problems of an antenna site and tower would drastically reduce the capital and operating costs for this part of the station and save money for programming, another example of the SeventyNine philosophy in action. Cable would gradually serve as the "equalizer" between VHF and UHF.

"Big name Hollywood-type programming," similarly, was out. "More low-key mix of information and entertainment with a local flavour" was in. The licence application had a weighty little essay entitled "What is Community," ending, with capital letters: "ULTRA HIGH FREQUENCY ... CAN ONLY ENGENDER A MEDIUM FAR MORE FLEXIBLE, FAR MORE DEMOCRATIC, FAR MORE DIVERSIFIED AND RESPONSIVE TO PRESS-ING NEEDS." SeventyNine estimated that it would be "able to achieve 80 per cent or more of its regular program schedule through in-house and in-town production."

Widespread, independent community ownership of the station was another hallmark. The station would be "free from media and economic allegiances which more often than not stifle initiative and innovation." Advertising would be low-cost and tap the local retailers market for whom the big VHF stations were too expensive.

After the station went into operation, Znaimer told a business-men's luncheon that "we have three production techniques: We have poverty; we have inexperience, and we have fear." "The little giant of Toronto television" was how the station was soon to be described by obliging Toronto *Star* television writer Jack Miller.

The principals of Channel SeventyNine Limited included Phyllis Switzer, the wife of a broadcast engineer, and Znaimer, an ex-CBC producer, both of whom were vice-presidents. Chairman was for-mer Loblaws head Leon Weinstein. The president was Ben Web-ster, a tycoon friend of Znaimer's. Jerry Grafstein, the Liberal operative and broadcasting lawyer, was also a founding director and secretary. Alan Eagleson was a director. Bobby Orr Enterprises was a shareholder. Newspaper columnist Ron Haggart was a share-holder and became chief of the City-TV news package. Singer Syl-via Tyson was also a shareholder. Switzer, who had once done public relations for the Canadian Cable Television Association, orig-inated the idea of the UHF station. Znaimer, who became managing director and was not yet thirty when the station went on the air, most visibly represented the station.

About a year after City-TV went on the air, journalist Val Clery did a profile of Znaimer in *Maclean's*, "Everything's coming up Moses: How to succeed in the television business with Moses Znaimer really trying."

Clery traced Znaimer's short and impertinent life as the "youthful boy wonder/*bête noire* of the CBC" — three years of rapid rise to television producer and sometime on-screen personality, ending in his resignation when a program format he had been developing was turned down by his superiors. The event everybody remembers, though, was an article in the *Star Weekly* magazine headed "Ladies and Gentlemen, *Meet* the Next President of the CBC," namely the young and zappy Znaimer who allowed that "under certain circumstances, if certain things about the job were changed, sure, I'd want [the job]." This was Znaimer's undoing at the CBC, Clery explained. Znaimer's effrontery was "a direct threat" to the mediocrat's belief in not rocking the boat as the means to the highest office of the corporation. Or so Clery said.

Over at City-TV, by invidious comparison, there was an aura of revolution. Or at least the events were seen by Znaimer in terms of revolutionary metaphor. Znaimer, who had an M.A. from Harvard in Soviet Studies, told Clery that "I'm . . . very conscious of the history of revolutions, so I'm determined not to see either this station or myself become what is often happily and emotionally accepted as noble failure. And I'm going to do my damndest to survive the transition period that distinguishes Che Guevara from Fidel Castro, from the type of person who starts things and the radically different type who keeps them going. I'm going to be both kinds of persons."

"If the clean-up men, the counter-revolutionaries, ever do have to move in on City-TV," obliged Clery, "they will not find in Moses a leader corrupted by absolute power, but a man who has run too fast and too far for ordinary men to keep up with."

City-TV began with a two-and-a-half-hour prime-time nightly news package beginning at 7:30 P.M., in the heart of peak viewing hours. Weekends were heavy with ethnic programming. A ring was set up in the studio for Saturday night boxing. One of City-TV's most imaginative innovations was *Free For All,* first scheduled late Sunday, which invited passers-by in off the street to say what was on their minds. *Free For All* was to generate its greatest publicity when members of the fascist Western Guard broke up the show one week during an item featuring black musicians. City-TV's great success, however, and a crucial source of revenue, was a soft-core Friday-midnight *Baby Blue Movie,* not envisioned in the original licence application. Another key element was a regular weeknight movie beginning at 10 P.M. and over before midnight. This was not indicated in the original application, either.

By the start of the second season, though, the revolution that never was a revolution was already over, except for the Cinderella image. The news package, previously cut down to two hours, was moved to 9:30 P.M., on its way out of the peak viewing hours, to see if it could pick up a bigger audience. The idea of a drama department had been abandoned almost from the beginning; a director hired on retainer was let go only a few months after broadcasting began. Revenues for the first year were more than double the projection, but so were costs. Although the first-year deficit was in line with the original blueprint, City-TV, according to reports, was eager to break even in its second year rather than in the fourth year, as projected in the licence application.

The hunt went out for imports with greater drawing power, mostly from Great Britain, whose programming was priced lower. The sexy soap opera from Australia, *Number 96,* was brought in to add to the Friday *Baby Blue Movie.* (The very first imported program on City-TV, shown the first night, was the controversial BBC series *Casanova.*) *The Price Is Right* was brought in for one night a week. *The Mike Douglas Show* was scheduled to run 4:30–6 weekday afternoons, the same time it was carried by a Buffalo station and therefore eligible for simulcasting on cable. *Perry Mason* and *Let's Make a Deal* were also acquired.

After its second season, City-TV applied to move its transmitter to the CN Tower. The CRTC obliged. There went the idea of a UHF transmitter located on the roof of a high-rise and its exemplification of the small-is-beautiful philosophy which had been peddled with such gusto in the application. But the pretentious rhetoric about the UHF style had always been as phony as it was fashionable; UHF was just a technical matter of transmission. The audience reach rose from 2.7 million to 3.7 million. City-TV wanted to get into Hamilton, Oshawa, Whitby and other areas by cable carriage and through its new boosted power. "Local" was not as local as it used to be. (Znaimer in 1971 had argued that "if the physical plant that you have is a hedge in the direction of regional expansion . . . sooner or later . . . you will abandon, or you will be reasonably distracted [from], your original *raison d'être.*")

No sooner was the transmitter approved in mid-1974 than new programming acquisitions were made. Reruns of *The FBI, Owen Marshall, Get Smart, Love American Style* and *Green Acres* came on stream. The British programming was out. The local information package was down to an hour and a half, though back in the peak viewing hours of 8–9:30 P.M., followed by half an hour of a world-news feed. The weeknight schedule from 5–11 P.M. consisted of four hours of American reruns, inside which was encapsulated the information package. *City Lights* with Brian Linehan came on at 11 P.M. The show's forte was interviews with Hollywood and other international movie and entertainment personalities.

"Yesterday, as he talked about [his 1974–75 schedule and turning a profit in only the third year], Moses faced the problem of not sounding too successful," wrote Jack Miller admiringly, "because his station has won a lot of sympathy as the struggling infant in Toronto television's land of giants. If the 'struggling' adjective became passé, he feared, maybe the sympathy would too. And so the wide-eyed plea that everyone go on loving CITY because, gee whiz, they're still smaller than the others, and they still try hard."

Znaimer and City-TV then parachuted an application into Vancouver when the delayed allocation of a new licence there came up in 1975, all the better to spin off some of City-TV's costs. Well practised by this time, Znaimer went through the old UHF routine with effervescence. The local opposition that developed he attacked as parochial, probably with all sincerity. Provincial and backward Vancouverites were to be grateful for having the UHF revolution brought to them by the ardent geniuses who had created it all in the first place. Presumably, too, the value of being "independent, community-owned" and of putting out a "purely community TV signal" — phrases used in getting the original City-TV licence — applied only if one started out in Toronto. Unfortunately, people in Vancouver did not realize that hustling local air-time sales with blue movies and reruns of *Green Acres* was spiritually akin to the works of Che Guevara. Phyllis Switzer disclosed later what the real purpose of the Vancouver application was: to relieve the heat in the City-TV boardroom generated by discontented shareholders.

Back in Toronto, City-TV's licence came up for renewal. By this time, early 1976, the baby blue movies had been dropped. "The Commission anticipates ... there will be no deviation from [the] objective" of the original application, said the renewal decision with the commission's usual authority. With this renewal safely in hand, City-TV then went on a deviation binge bigger and better than ever.

A month after the renewal decision, Znaimer was in Hollywood bidding for first-run U.S. network shows for the first time and buying stronger movies. A few months later, the new, new schedule (fall of 1976) was made public. The *City Show*, the local information package, now down to an hour, was put into the 6–7 P.M. slot just like the news programming of other stations. The key 8–9 P.M. slot was dubbed "the action hour" — second-run American series high

in violence like *The Rookies, The FBI, SWAT* and *Mannix.* From 9 to
11 P.M. Mondays, Tuesdays, Thursdays and Fridays, the station
scheduled movies, many of them also violent and many television
first-runs, including *Straw Dogs, Soldier Blue, Carnal Knowledge* and
Sleuth. The same slot Wednesdays was for imported specials. The
late afternoon, 4:30–6 weekdays, was also American. Paying for the
cost of the new transmitter and taking advantage of the larger
audience reach were the kinds of things that a trip to Los Angeles
was for.

Znaimer then appeared before the Ontario Royal Commission on
Violence in the Communications Industry, chaired by Judy
LaMarsh. Znaimer told the commission that City and others had to
buy American television shows despite the violence and public criticism. If they didn't, they would be wiped out. "I don't make the
violence. I've got to buy something. If I don't buy that I'll have
literally nothing to buy. If I don't buy it, the [advertising] agencies
stay away in droves . . . If I couldn't have *Mannix, The FBI, Cannon*
and *Mission Impossible,* what would I run? . . . The fundamental fact is
that the stuff is imported. If you want to change it, the government
has to create economic barriers. . . . If I have a reaction to the
fundamental waste of this commission, it's that you can't do anything." If Canadian stations had to rely on nonviolent material from
the United States, "all the Johnny-come-latelies, the third and fourth
stations in the market, [would] die economically in six months."

So much for the City-TV revolution! What could have made it
any clearer? Who would want it any clearer?

"I can't say it's been grand, but it's been stimulating," LaMarsh
sighed when Znaimer was finished.

The following summer, August 1976, City-TV bought a full page
in the *Broadcaster* to advertise its wares, that is, to tell advertisers just
how much like any other station it had become since 1972. Something like the pigs in *Animal Farm,* it had taken City-TV only three
years to join "the established hogs of the electronic frequencies"
with their "near uniformity of fare" and their "big-name Hollywood-type programming," so eloquently decried in the original
application. The "CITY-TV FACT SHEET" started with the transmitter
(1815 feet high), the enlarged "A" Contour Population and the
boosted transmitting power, and worked though the number of
cabled homes down to the programs. The first eight programs were
the American acquisitions listed in block capital letters, showing the
progress from 1972. The bottom three programs were the local
evening shows, in upper and lower case, looking like an after-

thought. "Now our schedule and coverage area compares well with other Toronto area television stations," the fact sheet proudly declared. And later, in a moment of convenient doubletalk, it announced: "We want to continue to provide vivid examples of how serving the public interest can result in exciting programs." Call it *Television Farm.*

The new programming philosophy and schedule was launched at an outdoor circus-style party, complete with baby elephant.

"It's part of growing up," said Znaimer. "I sort of regret that."

It was actually part of several City-TV shareholders getting edgy and putting on the heat. The previous winter, the station had reached $2.5 million in debt and was slipping deeper. It was called a "crisis of confidence." Znaimer was sent shopping for financial help and came up with Montreal-based Multiple Access, who would acquire 45 per cent of the stock and bail out the discontented shareholders. City and Multiple Access both piously said this was not control. The headline over the story in the paper, where they were quoted, said: "Montreal firm buys control . . ." You could call it what you wanted. Everybody got the picture. Multiple Access was the owner of CFCF, the Montreal CTV affiliate. It was controlled by the Charles Bronfman side of the Bronfman family, one of the good old "media and economic allegiances" which Channel SeventyNine had got its licence by denouncing. There went the rhetoric of independent community ownership (the reality of which had never existed; the station was not owned by the community, only by a group of shareholders).

For the hearing on the takeover in Toronto in December 1976, Znaimer and Switzer, no doubt with the help of Jerry Grafstein, concocted a soap opera of their own. The main character was poor City-TV, whose management, Switzer explained, had "insisted a hundred and ten per cent on delivery of our promises of performance" and had been abandoned by "the system." Disgruntled City shareholders had wondered why City had to be so virtuous when, looking at Global, they could see clearly that "the system itself didn't demand that." Discord and wrangling among the shareholders was the result. Management had then pursued the notion that a licence in Vancouver might be seen by the system as a just "reward" for those who were "doing things right" — licences in the colonies like British Columbia, obviously, were to be dispensed as patronage to the CRTC's Toronto clients. But they were disappointed there, too. Spiritually crushed City-TV was now so distraught by its discord — here Znaimer picked up the story — that it could not even meet new challenges, like additional licences elsewhere, real estate deals and pay-television (presumably these were the birthright of a meritorious television Cinderella which did things right). City-TV, as well, was all alone in a cruel world, not only when it came to sharing costs with other licensees but also "from the vantage of simple human needs; it's important to have someone to talk to and from whom to get advice from time to time." Relief, however, was at hand. It was nice to have Multiple Access "by our side," said Znaimer, looking to a new, more comforting day.

There was a dark villain, too, in this scenario, in case you hadn't guessed: Global Communications and mercenaries like Allan Slaight. "The new IWC/Global [IWC Communications took over Global in 1974] had survived intact by radically destabilizing the marketplace," charged Ms. Switzer indignantly. Global's wholesale shift of emphasis from Canadian to American had driven the price of foreign programming goods up and had also damaged, by implication, the reputation and hence the salability of Canadian programming goods, and hence the viability of City's original magnificent concept.

As luck would have it, the renewal of the Global licence was being heard during this same CRTC Toronto visit. City-TV filed for the occasion a sweeping intervention ripping Global apart. Slaight, however, mounted a counter-attack, pointing to City's own shucking of promises and sellout to American reruns, cop shows, sitcoms and movies, and to other real and alleged City sins. The pot should not call the kettle black.

It was a falling out among *Television Farm* manipulators, Orwellian style, scratching and backbiting. By the time they were through, they had done an entertaining job on each other and also, indirectly, on the commission, for having let the other guy get away with what the other guy had blatantly got away with. Nothing was lost, however. Since everybody, including the judge, was incriminated, everybody was safe. So Global's renewal came to pass.

One way a regulatory agency has of disciplining an errant licensee is to deny takeover applications. The CRTC approved the City takeover. "The commission," this particular decision concluded with the usual authority, "will accordingly expect early and significant steps to make City-TV in the heart of prime time a genuine alternative to other Toronto area stations, with that alternative to be based on Toronto produced and oriented material."

Multiple Access had been very obliging at the hearing. Once this decision was out, though, it threatened to call off the deal. The change outlined in the decision would make the station a perpetual money loser, it argued. So chairman Harry Boyle and the commission, after a little private meeting, conveniently let Multiple Access off the hook. "We were given relief on the ruling," Multiple Access told the Toronto *Star*.

Harry Boyle denied the commission had backed down. "Multiple Access," he explained, "has a lot of accountants in its head office and they tend to read documents literally." In other words, commission decisions do not mean what they say. The decision was not meant to be as harsh as it sounded at all, said Boyle. All the commission wanted was a "reasonable reversion" towards the kind of prime-time local programming City used to have.

The granting of "relief on the ruling" in effect constituted an amendment to the takeover decision. This should have been done with an application, a notice, interventions if any, a hearing if necessary and a properly recorded decision. What was a little sidestepping of public procedure, though, when all that was involved was the CRTC getting together with one of its family and saying its own decision wasn't so?

City-TV was now home free and winging, in "reasonable reversion" to its original mandate, bringing in such local Toronto programs as *NFL Monday Night Football,* a prime-time imported movie every night and twice on Fridays ("the big stars, the big titles, the hot releases . . . all the time" said the full-page sales pitch in the *Broadcaster*); and a "great daytime package" including *Joker's Wild, Match Game* and soaps like *As the World Turns* ("It comes down to one word. Audience. This year we're gonna have more of it. Watch the BBM's [the ratings] and grow with us."). Just to make sure everybody got the message, this jump into the big-money league was announced in a ballroom at the Hotel Toronto rather than in the board room at the station. A Multiple Access veteran moved in to reorganize the news show. The movies were flogged heavily in advertising. And Al Bruner, no less, the guy who dreamed up Global which City had been so bitterly attacking, was hired. Bruner was given a one-year contract as a consultant to "direct the market development of the . . . station," namely to help make City less like any relic of the previous City and more like a Global mutant.

There they were, Znaimer and Bruner, shaking hands and discussing in the board room, for the benefit of the press, how they would work together to develop new packages to attract the big sponsors. The station was ready for the plunge into profits, Znaimer said. City was "moving into the mainstream," Bruner said.

"Al didn't just have one good idea in Global, he's a man who's having great ideas all the time," Znaimer said. The new chairman of the board, a Multiple Access man, standing by, did not say much, but he wasn't complaining. As the press described it, the deal was one of the industry's great let-bygones-be-bygones gestures.

It wasn't long after, in early 1978, that City-TV announced it was going to simulcast the CBS opinion and public-affairs show, *60 Minutes*. Simulcasting on cable would add to City-TV that part of the Toronto audience watching the show on the Buffalo channel.

The unspoken general assumption was that purchasing U.S. entertainment programs was one thing; but public-affairs and opinion programming, well, well, if we let Americans do that kind of programming for us, we might as well throw in the towel. Unlike the high cost of entertainment programs (situation comedies, adventure drama, variety), public affairs of one kind or another was within the financial reach of Canadian television. Public affairs was something Canadians could do, something we were good at. Our broadcasting structure would hold the fort there.

Ex-CRTC chairman Harry Boyle, who had casually let City-TV off the hook less than a year before — Pierre Camu had since become chairman — was all of a sudden furious about the *60 Minutes* simulcast. "The whole industry here seemed to give up on Canadian entertainment a long time ago," Boyle lamented, "but at least we managed to keep a hold of our own information programming in TV." He called the City-TV move a "bloody crime ... especially after all the promises about local Toronto emphasis they made when the CFCF people bought in, with all their Montreal money."

The righteous indignation was just a bit late.

Global Television, which could outcrass City-TV any day of the week by virtue of its larger transmission reach, also knew a good thing when it saw one, or when somebody else saw one first; deftly, two years later, Global took Canadian rights to *60 Minutes* away.

City-TV meanwhile, in 1978, was in financial trouble again, its debts passing $4 million. Multiple Access wanted out of its 45 per cent holding. The "founders," as the original nucleus of shareholders was called, then struck a deal with Alan Waters of CHUM Ltd., giving CHUM 67 per cent ownership to add to its multiple television, radio and other holdings. The City-TV "founders" would still play a major role in programming. CHUM, of course, as the CRTC decision classically put it, "endorsed and committed itself unequivocally to the original City-TV programming and management concept." The station, Znaimer said later, with a reported satisfied sigh, "has largely returned to its original design, intensely local news and movies in prime time." Actually, the station's schedule looked nothing like the original design. The proposed program-blocking in the 1971 application envisioned 70 per cent local content and 80 per cent Canadian content in prime time, with two hours of locally produced drama weekly, local sports instead of NFL football, a single-edition two-and-a-half hour nightly information segment, and its main movie block running after midnight, out of prime time. But Znaimer, the UHF kid, was such a believer in his powers of freestyle imaginative conceptualization that he might have said anything.

Maybe the best part of the CHUM deal was Waters's offer to help the "founders" financially and to bring their shareholding from 19 per cent, to which it had fallen, back up towards the original 30 per cent This would be done by buying some of their shares for cash, then selling them back with five-year interest-free loans to cover the repurchase. With a gift like that, the founding shareholders were not likely to object to concentration of ownership and opt for another buyer who might not be so forthcoming. As usual, the only applicant allowed by the CRTC for the public licence was the one hand-picked by the outgoing controlling group. Alan Waters was home free. "A great partner," Znaimer called Waters.

In the spring of 1978 City-TV, along with four other stations, was charged by the CRTC with falling short of the required Canadian content in prime time, and summoned to court. "I'm not terribly nervous," Znaimer said. He would, he said, "pursue a very active defence." History, alas, never got the chance to record the defence. Lawyer Grafstein got them off the hook because of a mix-up over court dates.

8

CITV Edmonton
HOW TO PLAY THE GAME

SPECTACULAR AS THE Global fiasco was, ridiculous as the endless self-stroking, self-preening prattle of City-TV was, nothing quite beat CITV in Edmonton for a lesson in (a) how to construct an application for a licence, or (b) how to abandon the promises of the application. You could learn from the application, for example, that Van's Sausage Co. (no relationship to any of the CITV principals) spent $100,000 on construction in the last year of record (it was 1971) or that there were no winter closures on the Yellowhead route. There was lots more where that came from. The CITV (Edmonton Video Ltd.) submission, in the spring of 1973, was, in short, a textbook application, just what one might expect from Alberta up-and-comers in the big leagues, outdoing in professionalism and slickness, if not in philosophical concoction, Toronto amateurs.

Here is how it is done, in a few easy steps:

1. Start with one or more local capitalists with stature, yourself if possible (*Dr. Charles Allard, a surgeon turned developer; Allarco Developments Ltd.*).

2. Involve as a minority interest an existing media company, while also expressing your readiness to establish the undertaking without their involvement (*Selkirk Holdings, a television-cable-radio conglomerate*). The patina of professionalism and corporate substance associated with the media company will not be wasted either way.

3. Add a major entertainment personality and freely accommodate this personality in the schedule and in programming ideas (*Tommy Banks*). This personality will make an eloquent plea for doing television production, including prime-time programs, in the locality, and will explain how the pioneering spirit of the area will make it happen.

4. The entertainment personality may come first and local capitalists of stature may come second or third. In such a case, the local capitalists of stature will take control one or two years down the road instead of from day one. Mix and match.
5. Insert program promises. This is the least difficult of steps. The first all-purpose rule is to promise to do, and to be, something different than existing stations. The second all-purpose rule is to proclaim freedom from the obsession of maximizing audiences. Do not stint on the sincerity, no matter how mawkish. That's all there is to it.
6. Accumulate "letters of support" by contacting all and sundry, up to the premier of the province. Include all letters that come back, except denunciations, under the application category, "Letters of Recommendation from Civic Officials, Individuals or Community Organizations." Most people are polite enough to write something, particularly on an impressive-sounding notion like a new television station.
7. Mine this collection of letters for useful excerpts. Also, at the hearing, refer to the whole collection, in a casual, unspecified way, as "endorsements" from a broad spectrum of the community and "strong support."
8. Add economic information. This information does not have to be read as long as it is there. The technique is impressionistic. Great volume can therefore be used, and should be used. Overcome any misgivings that piling on so much superfluous information is ridiculous. This economic profile can demonstrate at one and the same time that the commercial market is sufficient to finance the promised programming and that the applicant has its feet on the ground.

Imagine that you are a member of the CRTC and are perusing the Edmonton Video application. Your eye passes over the pages. ... "Canada's First Organ Transplantation Research Group at University of Alberta" (Edmonton *Journal*); "Edmonton Just Keeps Growing" (*Canadian Banker*); Population Forecast, Population of City of Edmonton by Age and Sex; Births, Deaths and Migration, City of Edmonton; Births, Deaths and Marriages per 1000 Population, City of Edmonton; Ethnic Background; Population by Place of Birth; Taxation Statistics; Personal Disposable Income; Alberta Population Analysis; Edmonton Trading Area; Projected Total Population for Selected Census Metropolitan Areas; Building Permit Comparison, Metro Edmonton and Province of Manitoba; Value of Building

Permits 1968–1971 for Prairie Provinces; Total Value of Construction Work Performed for Prairie Provinces; Breakdown of Prairie Province Construction Expenditures; Value of Shipments of Own Manufacture 1970, Prairie Provinces; Western Canada Shopping Centre Comparison; Estimated Total Restaurant Sales, Prairie Provinces; Registrations of Passenger Vehicles, Prairie Provinces; Metro Edmonton to Pass Metro Winnipeg in Population by 1976; 1971 Age Breakdown for Metro Edmonton; reprints from the *Globe and Mail, Wall Street Journal, Texas Oil Journal;* Analysis and Summary of Investment in Industrial Projects, Province of Alberta; University Full-Time Enrollment in Regular Winter Session; Edmonton Visitors Bureau, Visitor Registrations, Estimated Visitors; Summary of Edmonton Conventions 1965–1971, Including Number of Conventions, Number of Delegates, Money Spent by Delegates; Industrial Projects 1971, Edmonton; also Industrial Projects, Northern Region, Calgary and Southern Region (Giuseppi's Pizza Ltd. spent $300,000 plus in industrial projects in 1971, Van's Sausage Co. spent $100,000 plus, Imperial Oil had a projected refinery expenditure of $200 million).

The *pièce de résistance* is the detailed comparison of Trans-Canada and Yellowhead Interprovincial Highways, including mileages, population, elevations and grades ("only at four points throughout are any major uphill-downhill sections encountered"), and climate (including summary winter closures).

9. In answering the last question of the application ("applicant's conception of public service broadcasting"), do so at sufficient length, no matter how long it takes and how vacuous the answer.

Here is a sampler of Edmonton Video Limited's "CONCEPT OF PUBLIC SERVICE BROADCASTING":

This question prompted a great deal of discussion among the parties to this application. Whatever initial temptation there was to escape into exposition of appealing philosophy and fine writing was soon dropped in favour of an attempt to say what we believe and can live by should we be entrusted with the responsibility we seek. [Translation: For lack of any good ideas, we are resorting to an exposition of appealing philosophy and fine writing together with a pro forma denial of doing so.]

Public service is not a separate component of television station operation, a matter of free air time for worthy causes, but is, rather, the essence of the entire operation. The total activity of a television station must be conducted in a manner that reflects the purpose of serving the public and in a manner which will be beneficial to them. Public service *is* the function of the station. All else is secondary. The station must, for example, operate at a profit over the term, otherwise it could not properly serve the public. But it is still a secondary imperative.

Our trust, ensuing from the grant of the right to use public property, is to serve the people of Edmonton. [Translation: Public service is service to the public.] How do we serve them? As discussed in Schedule 21, we intend to provide programming designed to fulfil presently unmet needs and interests of Edmontonians. . . .

Good character and performance are not enough. It is imperative as well that we properly comprehend our role in the Edmonton community. This is a complex matter in practical realization. In a social sense we must be truly reflective of that particular community; Edmonton is unique as is every community. The station must be an Edmonton station; the men and women of the station will know, and feel, that they are Edmonton people, that their job is to serve the people of their community; and that commandment is their prime commandment. [Translation: This station in Edmonton will be chasing an Edmonton audience.] . . .

Finally, we must admit to the inadequacies of this statement, which are not the result of deficiencies in phrase-making, but from a search we have yet to complete for the ultimate and encompassing beliefs about our responsibilities to the public. We shall continue on this quest. [Translation: This effort at fine writing is a pious crock, so we are saying so, in a few ultimate and encompassing words, to cover our ass.]

Having followed these few easy steps with well-heeled slickness, Edmonton Video Ltd. was awarded the licence. Of course they deserved it.

With the licence in hand, CITV, as the new station was called, had to decide what to do with it. "It's dead already," said one local TV executive, several months before CITV went on the air.

This terribly cynical comment was based on the terribly cynical assumption that CITV would attempt to fulfil all its programming promises for a few months or a year, and then, running into financial difficulty, would abandon its programming promises, just as Global and City-TV were about to do, and as all the CTV stations had done before them. CITV, however, had a better idea that out-Globalled Global. It began to sidetrack a few of its more troublesome programming promises before going on the air. It was pointless to lose money going through a futile demonstration exercise showing that the programming was commercially impractical or that its production overextended resources. Similarly, it made eminent sense to go directly from promise of performance to abandonment of promise of performance without the complications of passing through an opening schedule kept to the mark for the sole purpose of establishing brownie points with the CRTC. The commission could be looked after later.

One of the first commitments to be cut back was the weeknight public-affairs show originally scheduled for 9:30–10:30 P.M. It was reduced to half an hour before broadcasting began. U.S. entertainment purchases were thrown into the breach.

Insight, as the program was called, was supposed to be different from news broadcasts on the local stations of the CBC and CTV. It would, according to the application, be "more in depth than traditional news items of one or two minutes." It would concentrate on "reaction and analysis," making the most of brilliant specialists outside journalism and bypassing for the most part generalist reporters whose knowledge of any particular field would be superficial. The program would "move ahead of the traditional newscast route." All subjects, said a lyrical 1974 news release when the station was about to start broadcasting, "will be examined fairly and non-editorially, resulting in a penetrating search for an honest conclusion."

"We are capitalists and will not be afraid to admit it," said station general manager Wendell Wilks, explaining penetrating searches and honest conclusions CITV style. "We want to become the voice of business."

"Wall-to-wall boredom," wrote Edmonton *Journal* television critic Jim Davies of the program, "a half-hour piece of puffery that comes across as a series of advertisements for local personalities and organizations."

"What The Doctor [Dr. Charles Allard, CITV's owner and president] wants," one of the station's countless evictees told *Edmonton Report,* "is a program that offends nobody and interests everybody ... There isn't any such thing, but he won't believe it." The station kept firing its *Insight* producers, three in one year. Then *Insight* itself was canned, to be replaced with none other than *ITV News* to be run in none other than the 6 P.M. time slot, just when the other stations ran their news.

Programs on city and provincial affairs were held off until the second year. The six original Canadian dramas per year never made it on the air. The "regular English-language programming featuring French-Canadian artists" turned out, in practice, to be a purchase from a Sherbrooke station of a country music program which CITV eventually dropped for lack of audience and did not bother replacing. CITV had originally promised, "we will create such a program [featuring French-Canadian artists]." This showed the correct bilingual and bicultural ideological line and curried favour with the CRTC and other powers in Ottawa who were obsessed with it. Now CITV denied that creating such a program was what they had meant; they were only in the market for buying a program produced in Quebec.

The "six full symphony concerts per year" featuring the Edmonton Symphony Orchestra turned out in practice to be not symphony concerts at all, but largely American or other foreign entertainment programming, featuring performers like Leslie Uggams, Jack Jones, Engelbert Humperdinck, Vikki Carr, Tom Jones, Paul Williams, Dionne Warwicke, Wayne Newton, Tony Bennett and the like, with the symphony orchestra providing the back-up. These were dubbed "celebrity concerts" and flogged in the United States and other foreign markets for which they were produced.

The much-ballyhooed mobile unit spent most of its time hooked up to the studio like a technological gargoyle.

The key program in the application presentation and in the original proposed schedule was the *Tommy Banks Show*. For producing this and other programs, Wilks brought in people from Dallas, Portland, Montreal and a very heavy dose from Toronto. These imports were not exactly "the men and women of the station [who] will know, and feel, that they are Edmonton people," but what the heck, the *Tommy Banks Show* was not an Edmonton show, nor a western Canadian show, nor a Canadian show, and was never purported to be, even in the original application. It was just meant to be "Canadian content," an altogether different species (the celebrity concerts were also, of course, "Canadian content"). The *Tommy Banks Show*, like the concerts, was a backdrop for guests of allegedly "international stature" (read "mostly American"). Canadians were then thrown in around them. The show's existence depended on outside sales.

"It'll become a dead duck in one season," said a cynical Edmonton industry veteran about the *Tommy Banks Show*, thinking it would be a show for encouraging local talent. The show, with guests of "international stature," became a dead duck in one season anyway.

These shenanigans aside, CITV was basically an American station to begin with. In the first-year schedule, U.S. serials began at 3 P.M., leading into the regular weeknight movie (except for NFL football

Mondays in the fall), followed usually by another American serial. By the spring, the *Tommy Banks Show* was pushed from an 8:30 P.M. start to 9:30 P.M. (later it ended up at 11 P.M.). From 3–9:30 week-nights, CITV was thereby solid-block American and, like City-TV in Toronto, basically a station of imported movies. By running twelve hours of movies one week and sixteen hours the next in a two-week rating period, CITV at the end of its first year became Edmonton's number one station in peak evening hours, outdrawing the two older stations by far.

"Profit . . . is . . . a secondary imperative," said the CITV application before the licence was granted. "We're in this to make money," said Wendell Wilks after the licence was granted.

"We doubt very much that we shall ever proclaim ourselves 'Number One' in size of audience; such a slogan is alien to our concept," said the application before the licence was granted. "I think we can be number one in Edmonton," said Wendell Wilks after the licence was granted.

The presence of Selkirk Holdings "made our application much stronger in terms of experience and expertise," said Dr. Allard before the licence was granted. "The application was largely put together with the experience of Selkirk Holdings and, in looking back on it, I think there are many things that weren't realistic in that application," said Dr. Allard after the license was granted.

Maclean's, in the expert tradition of Toronto "national" journalism, ran a glowing article about CITV by Philip Marchand. Called "The Little Station That Knew It Could," the article generously excused the creation of an American production centre in Edmonton in the fictional guise of a Canadian station.

When the licence of "the little station that knew it could" came up for renewal in 1977, three years after operations began, the CRTC sternly told CITV that the discrepancies between promises made and actual performances were "a serious matter," particularly since the licence was awarded in competition with four other applicants on the basis of those promises. The commission renewed the licence for only eighteen months and ordered fulfilment within a year of the commitments made in the original application. This was later described as "a very tough decision" by a CRTC commissioner, Roy Faibish, who liked to indulge in the idea that the commission made tough decisions.

The second appointed time of reckoning duly rolled around, in 1979. CITV had corrected the minor discrepancies, like producing six dramas and the show for the native community (this latter conveniently scheduled at 11 P.M. Tuesday). But its block American schedule, from 4–10 weeknights, was largely intact, with an imported movie in peak hours Sunday and a movie after the CBC hockey game Saturday, to round off the strategy.

Allard and company had promised "a Canadian and local alternative . . . during the hours [between 8 and 10 P.M.] that [Edmontonians] most want to watch television." They were to draw on the energy of Edmonton's indomitable pioneering spirit and the rich Alberta talent pool. The subsequent CITV schedule was near perfection as the opposite. "CITV is really an American TV station with just a few little things that are just somewhat different," commented commissioner Jacques Hébert tartly. Hébert had previously expressed to CITV his frustration about listening to grandiloquent promises of performance and then coming back for a renewal hearing and learning that most of those promises had not been implemented.

But how was CITV's schedule different from other commercial Canadian stations, Allard asked. Hébert had to admit that it wasn't. It was "just as American as the rest of them," and, by inference, they were just as American as CITV. The feeble commission had licensed a new string of American stations, just as the Board of Broadcast Governors before it had done. Hébert, however, did not think of doing anything quite so logical as standing up and offering his resignation.

Allard then offered that television and movie production was like a chemical plant his company was associated with: "a world-scale plant but we couldn't have it in Canada without the American market." The schedule on which basis the licence was granted was "the wrong way to go." He, Allard, however, was innocent, having been taken in, he suggested, just like the commission. All the trouble had been caused by the bad advice of Selkirk Broadcasting whose personnel had drafted the initial application brief — pointing the finger at Selkirk was one of Allard's favourite gambits — although if the commission wanted CITV to bring back a Tommy Banks-style show in peak hours, he was quite prepared to do it, being a man of honour. Meanwhile, commissioner Roy Faibish was playing regulatory agency with such fervour that Allard at one point thought such a horrible instruction was actually going to come about, until Faibish reassured him. "You had me fooled, Mr. Faibish," joked Allard.

Faibish and commissioner Paul Klingle from Alberta elaborately went through the ritual of putting the various commitments, conditions and promises on the record and putting CITV's current programming on the record, step by step, going from one time slot to another. And then, with the help of Hébert and the other commissioners, they all renewed the licence.

CITV, or ITV to use its broadcast name, was never obliged by the CRTC to recreate the schedule for which it was licensed, or anything remotely like it. The station was put in the safe hands of a young accountant that Allard had brought with him into the operation. Creative people are not necessarily business-oriented, the doctor said, in explaining the new regime.

"CITV . . . symbol of what the West is all about . . . free enterprise . . . freedom from the tight reins of bureaucracy . . . success," purred the *Broadcaster*. "CITV . . . U.S. block programming from 7–10 P.M. weeknights . . . a Los Angeles satellite with hockey games," purred the schedule.

9

CKND Winnipeg
FOLLOWING THE FORM

THE CASE OF CKND WINNIPEG, the next station in the third-service licensing round, owned by CanWest Broadcasting Ltd., had its own distinctive wrinkles.

Like Edmonton Video, CanWest had religiously followed the rules for application packaging, including the patriotic invocation of sons and daughters who had in the past left their province for lack of outlets for their talent and had made it big in show business and television elsewhere. Unlike Edmonton Video, however, and Global, and Channel SeventyNine (City-TV) and all the previous applications from CFTO to BCTV, the original schedule proposed by CanWest did not even pretend it would carry Canadian programming in peak viewing hours. It was basically American to start with — U.S. situation comedy from 5 P.M.–6 P.M., leading into the news on weekdays, which was then followed by movies and more U.S. material from 7 P.M. through to 10 and 10:30 P.M. On Sunday night there were U.S. acquisitions and two movies back to back, from 6:30–11 P.M., and Saturday night, up against hockey, was "written off literally," as a CanWest shareholder subsequently let slip. The most cultural of the local programs, on which basis CanWest had solicited local endorsements, were dumped into this Saturday night hole.

In such a situation, where little or no alternative is offered, the proper strategy of the applicant is to claim boldly and unequivocally what a great alternative it is, which is what Paul Morton, head of the CanWest syndicate, did. The schedule was "designed to fill the existing gaps in present viewing options," Morton said. It would "fill a vacuum." It would "give prime time viewing to things which are rarely seen in Manitoba in those hours." And, pointblank: "We

believe to a very significant degree, it is an alternative schedule."
This was too much for even a captive CRTC to swallow, as their
skeptical but futile questioning on the point indicated. They issued
the licence anyway.

The CKND licence should never have been handed out. If the
application had been filed before City-TV and particularly Global
were licensed — if, that is, it had been reviewed on its merits — it
would not have been given a hearing, much less a licence. It would
have been considered retrograde and an insult, a duplication of
what was already wrong with the structure. It had none of the
promises, fantasy as they were, that had induced the commission to
license Global and to impose the rest of the licensing round on
western Canada in the first place. When the CanWest application
was filed early in 1974, it referred glowingly to the unqualified
success of Global; by the time the application was heard only a few
months later, Global had gone belly up. This brilliant prognostica-
tion by the Winnipeg promoters, plus the disturbing example of
Global, did not hurt the prospects of the application.

The CRTC, having rescued Global to save face among other things,
was not about to refrain from issuing a Winnipeg licence simply
because its schedule, based on peak hours, was essentially American
and against everything the commission was supposed to stand for.
This was exactly what Global's future would depend on, too, whatever the fanciful talk. Rejecting the best of the Winnipeg applications on this ground would have been tantamount to admitting, just
after the deed was done, that the rescue of Global was also a mistake. As well, a Winnipeg licence was necessary to buy some Global
programming and help Global out. Also, Edmonton had been
licensed — what would happen to it? The commission's action was
like a chief of staff throwing more troops into a hopeless situation,
since so many soldiers had already been killed. Not to do so would
mean admitting that the army command had sacrificed those
already dead not only in vain but also in gross error.

Similarly, the last thing the CRTC wanted was for any Winnipeg
licensee to follow Global into a messy imminent bankruptcy with
distraught creditors and bondholders, accusations of folly and
misjudgement volleyed left and right at the commission, a
bureaucratic uproar, an emergency hearing, and generally egg on
the commission's face. The Winnipeg situation, too, seemed a lot
weaker than Edmonton's, not to mention Toronto's. So the
licence had to go to the financially most stable organization. The
most stable organization was the one whose schedule was more

solidly based on economic reality, namely American programs in peak hours. So a management in harmony with this reality, that would not do anything foolish, was also a plus. Anybody who could argue that the CanWest application represented an "alternative schedule" — a baldfaced fiction — was just such a management. On the other hand, the co-operative applicant, Communications Winnipeg Co-op, could not be trusted. Aside from everything else, the co-operative had independent ideas about programming.

That is how the CRTC, while asking questions and making statements about alternative and local programming, issued a licence in Winnipeg to an essentially American station, this time knowingly, as part of a licensing round set in motion to offset American influence.

By the time of the first CKND renewal hearing three years later, in late 1977, the current programming schedule and the schedule in the original application looked, as one commissioner complained, "like two separate stations in two separate cities." Except for sports and the regular news slots, CKND had junked most of its provincial local programs, many of which had not got on the air at all, and replaced them largely with non-Manitoba Canadian purchases. That, plus an increase in broadcasting hours, meant that unrepeated local programming formed only a fifth of the schedule, rather than a half as boasted in the application presentation. These programming details were aside from the essential nature of the schedule — that its core, the peak viewing hours, were American. CKND was also packing more American content in the high-revenue fall period and balancing it off with more Canadian content in the low-viewing summer period — an old trick. Meanwhile, helped along by the rush to Los Angeles — the extra bidding of Global and then City-TV for American programs and movies for the Toronto market — the cost of U.S. acquisitions for Canadian television went through the roof. CKND's spending went with it. The prices CKND was paying were more than double what had been projected.

The commission, in its decision on the renewal application, authoritatively pointed out that the licensee had failed to produce many of the alternative programs it had promised and had also not fulfilled the requirement of scheduling some programs of a local nature in prime time. (These failed commitments would have been minor aspects of the Americanized schedule at best, although they were the only pretence available for the commission's having passed off the licence.) The commission also authoritatively drew to the licensee's attention that "this licence is granted on the basis of the particulars contained in the approved application." There were more authoritative words. The commission then lowered the local-program requirement (from thirty-three and a half hours weekly to twenty-four hours) and renewed the licence.

Of this whole cycle of licensing new third-service commercial stations, the only one which both promised a real alternative and at least retained the key format in the schedule was CKVU Vancouver, whose local *Vancouver Show* continued to run from 7–9 P.M., in the heart of prime time. The financial problems that ensued produced shareholders' squabbles and takeover schemes. Charles "The Doctor" Allard (CITV Edmonton) picked up shares and took an initial run at it. Izzy "The Flatlands Flash" Asper (Global and CKND Winnipeg) produced a loan to shore up the control of the major shareholders (Daryl Duke and Norman Klenman). The thick, complicated loan contract stirred up more contention and insinuations.

Even CKVU, however, otherwise followed the pattern, abandoning a promised daily outdoor magazine program and a daily consumers' program. Nor could the station resist resorting to traditional news and sports. It then extended its 5–6 P.M. news into the 6–7 P.M. slot, the same news slot as the other two Vancouver stations, and, with recycling and lots of U.S. news feed, making two hours of "Canadian content" out of one while it was at it. The CKVU licence application had included no news at all in its proposed schedule; the application in fact had gone out of its way to point out that the station would not carry news broadcasts.

The remaining successful applicant in the third-service licensing round was Télé Inter-Cité, which had been given the French-language licences for Montreal and Quebec City. It never got off the ground for financial reasons and surrendered its licences.

10

A Trail of Broken Promises

BY HANDING OUT LICENCES on the basis of certain promises and then backing off denying renewal when the promises were not fulfilled, the CRTC and the Board of Broadcast Governors between them created an Americanized broadcasting structure that was never intended and never envisaged in the legislation that the agencies were created to implement. Commercial television got into the house and took over most of the chairs in the kitchen under false auspices and with the strong-arming help of the American heavies they brought with them from across the street. This failed to stop CRTC chairmen from making nationalist speeches, as if the Americanized structure had just happened by stellar forces against which the CRTC was nobly battling, or at least over which it was wringing its hands at renewal hearings.

Anybody who could put two and two together could have figured out that commercial financing of television did not make sense for Canada. But the Diefenbaker and Pearson governments and the respective licensing agencies they created were either too ideologically captive and colonized to think straight about television, or they did not want to think straight about it. The worst mistake, though, was not the allocation of the commercial licences, it was the renewal of them, for then the demonstration of impracticality stared everybody in the face. If the commercial operators' actual performance had been known at the time of licensing, the licensing agency would have balked. Recreating the licences by renewal in effect created what the agency did not want.

The licences were, for all intents and purposes, American licences to begin with. "Canadian programming," a confidential CRTC research document pointed out, "is seen by the industry as the cost of a licence, not its function." Its function was to make money, which could best be done by exploiting cheaply acquired American programs, which is the opposite of what the licences were issued to do.

How do you regulate this antagonism? You cannot. "There is not a regulation that has ever been passed that someone cannot get around if they want to," Harry Boyle once told the Commons Committee on Broadcasting. Both Boyle and Juneau before him talked about how you could not regulate creativity and how co-operation was essential. They kept madly regulating, or pretending to regulate, anyway. Each time a royal commission or a government-appointed committee looked at the licensing and regulation of broadcasting, from the Fowler Commission and Fowler Committee to the Applebaum-Hébert Committee, it urged that the licensing agency get tough and deny renewal where promises weren't kept. In an allegedly liberal democratic society, licences awarded from the public domain to private parties should stand or fall by whether the licensees fulfil the promises on which the licences were granted. Anything else would be fraudulent. It would be stealing from the public. Denial of renewal for cause would also have focussed the minds of licensees. Handing out licences whose promises could be abandoned was also a disastrous way to conduct public policy.

Denying renewal would also have been instructive in ways not predicted by the various investigatory commissions. It would have wiped out commercial television in Canada, at least in English-speaking Canada — an ordinary, common-sense, businesslike step since its structure was impractical and perverse to begin with. After denial of renewal, nobody else would have bothered to apply. Take away pretence, evasion and abandonment and there goes the whole ball game. Canadians might then have begun developing new non-commercial financing for their broadcasting system, which did make Canadian sense and which could have done what the legislation said the system should be doing.

This process would have begun in the mid-1960s. By now we could have had a flourishing, sensible, economically efficient and productive Canadian television system, matching the British one, and putting up a real fight against the American invasion.

As it happened, instead of a Canadian broadcasting system, the Canadian Radio-Television Commission gave us Canadian renewal hearings. The old promise-of-performance trick, however, around which the hearings turned, was like a vaudeville act past its prime.

What was a poor, captive agency to do? This was proving embarrassing to the commission, so its machinery concocted something of clear bureaucratic genius: the request by the commission for "new promises of performance" which would allow licensees to write the original promises of performance out of existence.

The regulatory gavotte began with the Toronto-area stations which, the commission explained, "should not be weighed down by past commitments, which changing conditions have rendered obsolete." The new commercial Toronto-area stations had milked that very same excuse for all it was worth. They and the commission had both shied away from the much more relevant consideration that the expansion of television on the commercial side, even in Toronto, was an extension of the U.S. broadcasting and film production industry. This circumstance, in commercial operations, had been around since the 1920s. There was nothing new or unforeseen there. The new American-dependent licences, too, by increasing the competition for U.S. program acquisitions, and therefore the price, had made the structure pay that much more Canadian cash to the U.S. television industry. "A broadcasting environment has now been effectively constructed," explained the CRTC, trying not to pay attention, "which should permit the renewed commitment by licensees to the statutory objectives for broadcasting in Canada prescribed by Parliament."

Another marvellous thing about the new-promise-of-performance-because-of-changing-conditions caper was that it was detached from time. It could therefore be taken to any lengths. CKND in Winnipeg was not licensed until September 1974 and did not begin operations until the following August. Only two and a half years later, in March 1978, the commission dealt with CKND's failure to live up to its commitments simply by asking it to submit a new promise of performance. Everybody thought the "old" promise of performance was a new one. This was not, however, how broadcasting regulation worked.

What was required of the licensee, the commission explained, was "a renewal of its original programming commitment in realistic and achievable terms."

What if the new promises of performance were appreciably different from the ones on which the licences were originally granted — new licences, in other words? This is what "realistic and achievable terms" would work out to. Would the commission call for competing applications to allow those who were more realistic and honest the first time around, as well as newcomers, to apply?

What if the new promises of performance, despite pushing and shoving by the CRTC, consolidated what was already the case, namely a commercial television sector which, for practical purposes, was a branch of the American television industry? Would the commission resign in a body, in shame and mortification? Would even one commissioner resign as a matter of principle? No, no, there was such a thing as taking a "renewed commitment . . . to the statutory objectives" too far.

Meanwhile, as the new-promise-of-performance ritual slowly got underway, Harry Boyle resigned as CRTC chairman; or, as he was to explain later, he was given no choice by the Cabinet other than to resign, in the middle of his term.

To replace Boyle, Trudeau appointed Pierre Camu, the president of the Canadian Association of Broadcasters (CAB), the private broadcasters' lobby. Even case-hardened cynics did a double take. A number of unnamed "communication observers" contacted by Canadian Press allowed cautiously that it was unusual for the government to appoint the head of a lobby group to the agency that will regulate the lobby group's interests. Unusual, was it?

Editorial hell was raised. Michel Roy in Le Devoir wrote, "One must insist: It's not the good faith of Mr. Camu that is in question; it's the circumstances and the reality of things which, from the beginning, sully his impartiality and, in the eyes of the public, raise the conflict of interest." Geoffrey Stevens, the Ottawa columnist of the Globe and Mail, wrote that "no recent federal appointment has generated this immediate and angry censure that has greeted [this one] . . . To many people interested in broadcasting, it's a case of inviting the fox into the chicken coop." The Toronto Star commented: "What's so astounding is that Prime Minister Trudeau would be so insensitive to such a potential conflict . . . To some Canadians it's inevitable that Camu's decisions would seem unfair even if they were not. That's why the appointment is wrong and should be rescinded."

The Consumers' Association of Canada protested the naming of Camu. The appointment was also criticized by Pat Nowlan, Progressive Conservative communications critic, by the National Association of Broadcast Employees and Technicians (NABET), by the Federation of Communication Workers and by the National Anti-Poverty Organization, which had participated in a number of CRTC hearings. Ed Broadbent angrily rose in the House and asked for reconsideration of the appointment. The naming of Camu, he said, was extremely arguable. The precedent of Marshall Crowe, president of the National Energy Board, was raised. Crowe, by a legal action, was forced to step down from a panel hearing proposals for a northern gas pipeline because he had earlier been on the board of a company with an application before the panel.

"It could be worse, I suppose," a public-interest lawyer said sardonically. "They could have appointed the president of Bell Canada [the CRTC now also set Bell Canada rates] or a cablevision lobbyist."

Camu, however, could not see the problem, or pretended not to see it, which only confirmed the nature of the problem. For most of his life he had been a civil servant, including president of the St. Lawrence Seaway Authority. He also had a doctorate in political geography and had taught at university. Therefore, went the assumption, one could trust in his impartiality. It was a simple matter of changing masters. "After all," Camu explained to one reporter, "I didn't spend my whole life working for the CAB. It was only four years." In the next phrase, mind you, he did refer to his old clientele as "my broadcasters," but what's a little bit of natural human friendliness? The appointment did have, when one thought about it, an unbeatable historical logic. If an agency is captive to the industry, what better choice for its chairman than the representative of the industry? True, it did not represent exactly the "new wave" that Boyle had in mind for his successor, but rather the coming out of the closet of a relationship begun in the days of the Board of Broadcast Governors, a relationship of which Juneau and Boyle were pioneers. The fact that Trudeau and Minister of Communications Sauvé did not see the CAB presidency as an impediment to the appointment was a reflection of this state of affairs.

Camu had still more eminently suitable credentials. He was a product of the old Jesuit Collège Ste. Marie. A schoolmate who became a close and lifelong friend was Pierre Juneau. Another classmate, also to become a close lifelong friend, was Maurice Sauvé, a former Liberal Cabinet minister and husband of Jeanne Sauvé. Jeanne Sauvé had been first elected and joined the Cabinet in another portfolio in 1972. This is what had made Camu such a good find for the CAB four years earlier.

But why did Camu take the CRTC job, and at a salary cut to boot? Because, in case you could not imagine it, it provided him with the means, he said, to respond to the Quebec threat to national unity. "I think there are indirect means of contributing to the debate, the big

debate as we call it ... and I would like to help the cause of the
country." And later: "Certainly the Lévesque separatists have
learned to use [broadcasting] to project their image effectively."

So he had the right obsessions, too. This was particularly felici-
tous, especially after an inquiry into alleged separatist bias at Radio-
Canada, pushed on Harry Boyle and the commission by Trudeau,
discovered that there was no significant bias. (The inquiry had
issued its findings in July 1977. Camu was appointed to replace
Boyle shortly after.) Other than forcing broadcasters to become
propagandists for the federalist cause, or to tell their news people
to help federalists project their image effectively, or to correct their
separatist bias as apprehended by Camu, what "indirect means" did
Camu have in mind? Nobody asked.

For the hopeless CRTC, it was damnation and damnation again.
Global, which had boldly pledged in its 1971 application that its
news would "not be another repetitive and redundant 11 P.M. news
service," in the fall of 1978 moved its late edition from 10 P.M. to 11
P.M., where it would be just another repetitive and redundant 11 P.M.
news service. (Both *The National* and *CTV News* were then at 11 P.M.)
Global had been licensed as an alternative service, but the only
alternative it offered was its 10 P.M. newscast. Now it had gone and
done away with that, too, just at a time when its profits were rising
and it could afford to stay different. Worse still, it seemed, only a
year and a half earlier the CRTC had generously approved the complete takeover of Global by Global Ventures Western Ltd. in Winnipeg. This takeover approval, now of course irreversible, was
attached to the understanding that Ventures Western's "ethical
ownership," as the syndicate styled itself, would carry on with all of
Global's commitments.

Hurt feelings of betrayal promptly erupted, or at least were
reported to have erupted, in the corridors and within the dividing
partitions of the commission. Jack Miller wrote dramatically in the
Toronto *Star* that the staff's willingness to talk to reporters about
the incident "may be unprecedented in Canadian broadcasting history." A CRTC spokesman brushed aside Global's excuse that its 10
P.M. news time seldom drew half the viewers that the two Canadian
competitors drew at 11 P.M. The commission, he said, had bent a lot
of rules to let the new owners keep Global going after the 1974
breakdown. A main incentive for that leniency was the thought that
if Global survived, it could go on giving viewers alternatives in
program patterns.

Dr. Pierre Camu, as he was called, the man who by his own declaration set out to make the CRTC into the best regulatory agency in the world, was reduced to writing an angry personal letter to Global's president, Paul Morton. There had not been an application and hearing about the change — even a moronic licensee knew there did not have to be and these things were best finessed in midterm — so Camu's letter was just a useless one-man complaint. A CRTC source conceded it was too late for Global to change back now (though he did not say why it was too late; he also omitted to mention that the proposed change had been known to be in the works for some time). The source said Camu was simply venting his frustration. For that purpose, the letter worked extremely well. An angry personal letter is one of the best ways of venting frustration although, for a regulatory agency, a public hearing or a speech to a convention does not come far behind.

And then the Toronto area stations, which had been asked in a lengthy CRTC statement to file new promises of performance, simply refiled existing schedules instead. They did so in spite of the clarion call of the commission about "the tradition of Canadian broadcasting," and the "new challenges and opportunities" and the "strengthened economic capability" of the stations, and in spite of twelve closely worded pages of historical exegesis, stern lecturing and careful instructions of what was expected!

"Without exception," the CRTC found, "the television broadcasters in the Toronto-Hamilton area submitted Promises of Performance which reflected what is currently on the air or what was on the air in the spring of 1979 . . . The applicants generally did not respond to the comments and requests regarding future programming plans made in the Public Announcement . . . The commitments merely reflected the status quo." Upon which the CRTC asked for *newer* new promises of performance. This time, the new notice warned, they would be considered in connection with the licence-renewal applications of the stations. Their adequacy would be "carefully scrutinized."

Why any of them were expected voluntarily to offer to do more — and also indirectly to admit, thereby, that they could have been doing more in the past — was a question the CRTC apparently never thought of asking.

So there was another CRTC-licensee meeting, otherwise known as a public hearing, at the Park Plaza Hotel in Toronto, to consider the not-so-new promises of performance. This gave the commissioners an even better chance than writing personal letters did to vent their frustration. Chairman Camu recalled his anger over Global's decision to move its late newscast to 11 P.M. Jacques Hébert made his usual speech for English-language television stations, declaring that the schedule before him was not much different than an American one. Just before his statement, Global had used the old videotape caper, with newscaster Peter Trueman narrating. Trueman explained, among other things, how Global provided work for so many people, a lot of very good people, that they filled one parking lot (shown in all its splendour) and another one on the other side. Hébert's comments were pointed:

> I enjoyed very much your presentation this morning. . . . I couldn't help thinking that if that was a true reflection of the reality of the station, how great a station it would be. But you know as well as I do that it is not . . . When you look at the reality which is your program schedule . . . It looked to me so much like an American station that I was just thinking again what would happen if we showed this exact schedule to some city in the States, let's say Texas, Dallas, Texas, this exact schedule, except that we would change maybe the news a little bit. Not totally, because what is international comes from the U.S., or from U.S. sources anyway. So I have the impression that nobody in Dallas, Texas, would suspect that it was not an American station.

Hébert had been making these kinds of altogether useless declarations for several years now.

The angriest of the commissioners was Roy Faibish. Faibish was a former senior executive at Ottawa's privately owned station, CJOH. Before that, he had been part of the *This Week Has Seven Days* unit, and before that, in charge of programming research for the Fowler Committee.

"You guys certainly know how to get the blood of a mild-mannered man like myself boiling, and it's boiling at the moment," he said to a hushed crowd at the hotel. He referred to a previous commission decision that said, he explained, "that 'except for the news and public affairs, [Global] provides no alternative,' and that was a commitment to the commission and the public. And you've moved it. . . . And I say to you" — by this time he was thumping a finger on the table — "that you place your network licence in jeopardy against the background of that CRTC decision. I say no more. I am saying, you place your licence in jeopardy when you do that, given the fact that the essence of the Global Network was to be an alternative. That was the understanding. At least that's what the commission said. . . . I tell you, that when it's useful for a licensee to make arguments as to the virtues of having news at 10 o'clock . . . and at some point in time that is a useful argument to make because you want something [a renewal] from the commission, that that is rendered null and void . . . when you come back two years later to make arguments as to why it is a compelling . . . necessity to move it. . . . I tell you . . . the commission has heard enough over the years of that kind of stuff. . . . What is at stake at this hearing is really the credibility of the commission."

So the commission and not just Global was on trial — it was backed into a corner — and in that position, was the desperate suggestion, it was capable of doing some unprecedented things.

"Global TV told licence 'in jeopardy,'" ran the headline in the Toronto *Star*. The *Star* also ran a head-cut of Faibish. "Faibish: Shocks industry," said the cutline. Faibish's licence-in-jeopardy threat was, of course, altogether useless in inducing Global to switch its news back to 10 P.M.

Earlier in the day the commission had looked at the "new" promise of performance of CKCO in Kitchener, a CTV affiliate. The president of the station was also CTV chairman of the board. Faibish had exercised himself on that occasion, too. He warned that there was growing pressure on the CRTC to let other groups apply for existing broadcasting licences. "It's only a matter of time when other parties come along and say, 'Look, that guy helped build that company and he's had the licence for a quarter of a century and he's recovered his investment and made a little profit. I can do better than he can do now and I want a chance, and I'm prepared to take less profit and whatever the consequences are, and do more,'" he said. This was tough talk.

The previous fall, the CRTC had heard a renewal application by BCTV in Vancouver. Faibish suggested that if the financial details of the broadcasting industry were made public, which he wanted, there would be an outcry so strong that company licences might have to be revoked and the CRTC could be replaced by a new regulatory body. And if the CRTC were called before the public, it would be unable to defend the lack of spending by broadcasters for the production of network programs. This was provocative talk, too.

When the commission issued its announcement on the Toronto hearing, it scolded Global in general. It called the moving of the news to 11 P.M. "deeply regrettable." This fine exhibition of moral pressure was, of course, altogether useless.

Faibish at one point stopped attending commission briefings. The staff had nothing useful to say, in his view; he would read most of the evidence himself. At one hearing he sat in the audience and made audible disparaging remarks about the stupid questions the commission was asking. He was almost ranting, the proceeding was upsetting him so much. Later he began to develop the habit of occasionally asking long and rambling questions — miniature speeches — in an attempt to pin down the underlying anarchy of the process. Faibish just kept sputtering away. His arrangements to soothe his own conscience were great for gossip but, of course, in terms of changing the situation, were altogether useless.

Amidst all this sputtering was the CKO case. In the summer of 1976 the CRTC had issued a licence for a privately owned all-news radio network, later to become known as CKO radio. The CRTC issues hundreds upon hundreds of decisions per year, from major television-licence decisions to renewals of rebroadcasting stations in the hinterland. Each decision is preceded by a notice of hearing or application, also issued in a never-ending stream of paper. The poor monitor of the regulatory scene goes through this paper until his eyes glaze over. What, however, should the long-suffering reader of CRTC notices and decisions discover, in going through Decision CRTC 76-416 regarding CKO, but the following sentence: "Should the concept of All-News programming result in insufficient revenue to maintain the financial health of the service, the commission expects the licensee to surrender all the licences to the commission, rather than change programming in order to generate more revenue."

This was a most interesting and exceptional find. It was like coming across a nugget after panning through sand for half a lifetime. The Association for Public Broadcasting in British Columbia (APBBC), with almost illicit pleasure of anticipation, knowing the future but not knowing quite how the avoidance trick would be pulled, put Decision CRTC 76-416 in a special file.

The CKO network was the creation of Jerry Grafstein (the broadcasting lawyer, Liberal operative and City-TV founding shareholder) and Israel Switzer (cable and broadcasting engineer and husband of Phyllis Switzer, another City-TV founder). The other major participant was Agra Industries (Ben Torchinsky), the agrocorporation that had bought up several cable properties (for Weyburn, Estevan, Kamloops, Powell River, Chilliwack and Lethbridge). The all-news network's programming concept was the NEWSWHEEL (impressively spelled in capital letters by the promoters), an hourly package of hard news, actualities, backgrounders and short features turned over twenty times a day, every day of the week. The longest single segment was six and a half minutes. The all-news network, if it flew, would become "an important Canadian institution," Grafstein had opined before a credulous CRTC; the syndicate's NEWSWHEEL, meanwhile, had been designed by a consultant from WINS, a U.S. all-news station.

Sure enough, by the end of the first year, CKO was on the ropes. The first-year loss was several hundred thousand dollars. The second-year deficit was projected at $1.5 million. In order to save its skin, CKO began broadcasting Monday night NFL football (a great saviour for City-TV, too) and periodically other live sports events, such as harness racing from the U.S. Of such are Toronto "Canadian" institutions made. The station then bought the rights to Toronto Maple Leaf Hockey games out from underneath Foster Hewitt's declining radio station, CKFH. There went the famous twenty-hour continuous NEWSWHEEL.

The APBBC took Decision CRTC 76-416 out of its file and began work on a submission about the breach of promise. Foster Hewitt Broadcasting Ltd., trying to recover its hockey broadcasts, was the first to get something in. Hewitt, represented by none other than Chris Johnston, a former CRTC general counsel, asked the CRTC to order CKO to cease and desist. A hearing was called. Thanks to the special clause in the licensing decision, written in precisely for this kind of eventuality, the stage was set for the CRTC to defend its licensing integrity or, rather, unknown to Foster Hewitt, the stage was set for more comedy.

Poor Foster Hewitt! Unlike CKO's American-inspired NEWSWHEEL, Hewitt was truly a Canadian institution. Armed with a former CRTC counsel, he appealed for justice. The CRTC panel consisted of only three members, none of them the CRTC chairman or vice-chairman. One of them, the only part-time commissioner present, was Harry Bower from Regina. Bower, to a reporter from *Content* magazine, described Torchinsky, the president of Agra (CKO's principal shareholder) as "a very close friend." This connection did not induce Bower, however, to absent himself from the panel. Bower was also on the original panel that awarded CKO its licences.

Of course Foster Hewitt, jettisoned by Harold Ballard for a mess of CKO pottage, did not get his hockey games back. Of course the CRTC let CKO gently off the hook. But how did the commission get around their own key clause, specific to the licence award and upon which the whole matter turned, that "should the concept of All-News programming result in insufficient revenue to maintain the financial health of the service, the commission expects the licensee to surrender all the licences to the commission, rather than change programming in order to generate more revenue"? Well, they didn't mention it.

The APBBC, in a written submission from Vancouver, did focus on the clause, but the commission was under no obligation to pay attention to this representation.

Breach of promise? It was "all-news" to the CRTC.

11

CTV Drops
the Pretence

IN THIS SAME PERIOD of new-promise-of-performance games, the end of the 1970s, the CRTC considered the renewal application of CTV. The CTV network was long-established and national in coverage. Its major affiliates were prosperous and powerful. The owners included the Bassett and Eaton families (CFTO Toronto and CFQC Saskatoon, through Baton Broadcasting); the Argus Corporation (CJOH Ottawa, through Standard Broadcasting); Western Broadcasting and Selkirk Holdings, which included a Southam interest (BCTV Vancouver and Victoria); the Bronfman family (CFCF Montreal, through Multiple Access); Maclean-Hunter (CFCF Calgary); Alan Waters (CJCH Halifax, CJCB Sydney and CKCW Moncton, through CHUM Ltd.); Electrohome Ltd. (CKCO Kitchener, through Central Ontario Television); Moffatt Communications Ltd. (CKY Winnipeg); Harvard Investments, a real estate developer (CKCK Regina), and Sunwapta Broadcasting (CFRN Edmonton). Unlike Global, CTV, as an example of how commercial television worked in Canada, could not be passed off as an anomaly.

The CTV stations were methodically doing business as usual. They played the game with the CRTC in a similar well prepared fashion. Doing so was one of the things CTV president Murray Chercover was paid for. "Chercover," wrote *TV Guide* the week of the hearing, "can talk rings around almost anybody, and frequently does." CTV also had a special vice-president delegated full-time to "government relations."

This hearing revolved around "drama" programming, which put a premium on Chercover's ingenuity, since program production in that category was not on CTV's corporate agenda. "Drama" programs — situation comedy, adventure serials, or drama properly

speaking — were and are the core of television viewing habits and
the heart of any self-respecting network. They are also vital to any
country's self-expression. They are also more expensive than most
other programming to produce, which is why CTV avoided them.
For situation comedies and adventure and police programs, it could
get the Canadian rights to U.S. programs for as little as 3 per cent
of the production costs. There had already been a long song and
dance over the matter between the network and the agency — the
bashful suitor, the CRTC, trying to get the necessary co-operation
and submissiveness from the playful, seemingly affectionate but,
when pressed to do something, perversely willful and selfish CTV.
Declarations by CTV about the network's "very positive" attitude to
drama production and similar ardent protestations of virtue issued
forth at renewal hearings all the while. This routine went back to
the Juneau days in 1972.

Now CTV again was producing no regular weekly Canadian
"drama." The little that had been done in this category during the
three-year licence period consisted of international coproductions
with an eye to U.S. sales and no Canadian character to speak of.
Excuse My French, which was Canadian in context, had been
dropped in the previous licence period, despite good ratings,
because of a commercial deficit and little possibility of foreign sales,
although CTV stations overall were making more money than ever.
Canadian programs of all kinds scheduled by CTV during the peak
viewing hours of 8–10 P.M. had similarly fallen from two hours per
week to half an hour, out of a total of twelve hours (U.S. and
Canadian) supplied by the network in that time. Just before the
hearing, another half-hour Canadian was added to the time period
— an Americanized *Grand Old Country,* as it happened — as if that
were a great improvement. CTV, predictably, was now behaving just
the way an offshore branch of the U.S. commercial-television
industry should.

Meanwhile, pretax profits for private television, as a return on
net assets, were in the 55 per cent range. For the larger of the CTV
partners, like CFTO and BCTV, they were probably even higher,
although how much higher could not be pinned down, since their
financial returns to the CRTC were kept confidential.

The Council of Canadian Filmmakers, made up of ten trade and
professional associations in the English-Canadian film and television
industry, diligently filed a long and detailed intervention critically analyzing CTV and its history, laying out the whole dismal story
for the commission. The Association of Canadian Television and

Radio Artists (ACTRA) filed an analysis, as it had done in 1975, and in 1972, and in 1971, and in 1970. The editorial pages of the Toronto *Star* and the *Globe and Mail* even got into the act this time with attacks both on the games private television played with the commission and on the commission's feebleness.

The renewal hearing exercise began again, in all its pretentious banality. This particular episode took place in the Salle des Nations, Auberge de la Chaudière, the hotel in the new and massive block-square complex in Hull where the CRTC had been relocated. The hearing dragged on for three days. The questions and answers wound through convolution upon convolution. In the end Chercover and the rest of the CTV delegation simply stonewalled. This tactic would not have been used six years earlier, or even three years earlier. There would have been at least some suggestion of a co-operative attitude, however superficial. Now CTV could no longer be bothered to keep up the pretence. "My only regret at this hearing," offered vice-chairman Charles Dalfen, who had done most of the questioning, "is that that thread of commitment to substantially greater programming dollars does not come through from you. . . . It always facilitates [the] process when, as part of the ongoing interchange between us we get the commitment from you as well."

If the CRTC wanted a Canadian drama series (read "situation comedy") on CTV, without watering down the already compromised content requirement, it would have to order CTV to do it. Some of the "most persistent CRTC optimists," it was reported, were so disheartened by what had happened at the hearing that they could barely speak about it.

What a sad story. When the CRTC finally did order CTV to do at least some new drama (an average of three-quarters of an hour per week, of which half could be done in U.S. coproduction), CTV instituted legal action against the decision, resisting and delaying all the way to the Supreme Court of Canada.

12

Global Closes
the Circle

GLOBAL COMMUNICATIONS, MEANWHILE, did what came naturally. Along with the move of the late news from 10 to 11 P.M., which had so exercised CRTC commissioners, Global, in the fall of 1980, got a one-hour version of Johnny Carson's *Tonight Show* to fill the 10 P.M. slot. *The Tonight Show* (NBC), together with *60 Minutes* (CBS) — taken from City-TV earlier — plus liberal use of the ABC news feed (which did count as Canadian content), gave Global all the Canadian character it needed to make a success out of life.

At the beginning of 1980 Global issued a special announcement that it was "in the strongest programming position in its history." Four of the top six U.S. shows, it boasted, were in its schedule, along with four other successful U.S. shows. Of Canadian programming, only *Global News* was mentioned. This drove even the editor of the *Broadcaster* to comment about Global's preoccupation with American content.

There was another, still more deliciously ironic happening, one that closed the historical circle from 1972. But let us go back in history.

Global had been licensed by the CRTC in the context of broad strategic thinking (or what passed for broad strategic thinking) about the influx of American stations on Canadian airwaves. Juneau, the Rommel of broadcast border wars, matching intellect and training against superior forces, led the planning sessions. Sydney Newman once said admiringly of Juneau that he sat on the hottest seat in the country and had the coldest ass. Here was Juneau in action. The hopes for the strategy were so high and the drama attached to the campaign so affecting that the practical weaknesses of the Global application had been shunted aside.

The strategy had two prongs. One was to cut off Canadian advertising going to border U.S. television stations. This admitted of several tactical possibilities, among them the deletion of commercials from American stations on cable and the substitution of Canadian commercials. This idea was ultimately abandoned. Another tactic, simulcasting on cable (replacing the American station's signal with a Canadian one when the program was the same), which was adopted, did have a deletion and substitution effect, except that it was limited to those slots where simulcasting took place. Another possibility was tax legislation, encouraged by Juneau and eventually carried out by the government, in the form of Bill C-58, by which Canadian advertising on American stations could not be charged as a normal business expense. For businesses in the 50 per cent tax bracket, this would double the cost for advertising on border stations, everything else being equal.

The second prong of the general strategy was to license new stations in Canada that could be financed by, and could absorb, the repatriated advertising revenue. This was where Global came in

The main targets in the United States were the three major Buffalo stations available in Toronto and KVOS in Bellingham, Washington, a small city fifty-five miles from central Vancouver and thirty-eight miles as the crow flies from Victoria. These stations were not licensed to serve Canada. They did not have costly Canadian-content requirements. Their programming was American. Further, when they paid for programming rights, their Canadian market was a clear bonus. So these stations were extraordinarily lucrative. About 30 per cent of the advertising on Buffalo stations was Canadian.

The worst predator was Bellingham's KVOS, almost 90 per cent of whose revenue was Canadian. Without Canadian advertising, it would not exist. Its local news and public affairs were nominal. It was relatively the most profitable station in North America, with its program acquisition costs based on the local market, while its revenue came from the B.C. market four times the size. The president of KVOS was Dave Mintz. Mintz, by coincidence, had come to Bellingham from a radio station in Buffalo. In 1954 he, along with the KVOS engineer, had been responsible for coming up with the proposal to save the station by going for the B.C. market. The station moved its transmitter out to an island in the Strait of Juan de Fuca, gaining an unimpeded reach into Vancouver and Victoria.

The Canadian measures against the border stations touched off what came to be known as the "border war." The border stations,

KVOS among them, lobbied Washington to put pressure on Ottawa and hired Washington lawyers. Congressmen and senators took up the cause, most notably Senators Warren Magnuson and Henry Jackson from Washington State. The National Association of Broadcasters, the U.S. equivalent of the CAB, got involved. The State Department got involved. The U.S. consul in Vancouver got involved. Canadian-U.S. meetings were held. Talk of retaliation cropped up from time to time; preliminary measures were taken in that direction by ardent senators. Secretary of State Henry Kissinger brought the matter up in a meeting with Prime Minister Trudeau. Newspaper ink flowed. *Saturday Night* sent Morris Wolfe, their television columnist, from Toronto to Vancouver to take a look at KVOS and to interview Mintz.

"Canada does more than $40 billion in trade with the U.S. every year," said Mintz at one point. "Is it worth causing friction with such a partner for the sake of a mere $20 million [Canadian advertising on U.S. channels]? Let's negotiate." He offered $2 million for a Canadian production fund to be controlled in Canada if Bill C-58 were rescinded for the U.S. border stations. "We should not be permitted to operate in the way that we are operating," he volunteered generously after the step had been taken to stop him from operating in that way.

Commercial deletion and substitution (the replacement on Canadian cable of the U.S. station's commercials) would have closed KVOS. Simulcasting of programs, on the other hand, hardly affected it (it juggled its scheduling to avoid having the same programs at the same time as Canadian stations — counter-insurgency). To counteract Bill C-58, Mintz discounted KVOS's rates to Canadian advertisers, which, with its wide profit margin, KVOS could afford to do. In the summer of 1978, nevertheless, Mintz was complaining about Bill C-58 to the special trade office of U.S. president Jimmy Carter and asking for retaliatory measures that would make the Canadian government bend. The Vancouver stations, particularly CKVU, were still complaining about KVOS.

Not too long after, Global Television, which had been created as a key element in the campaign against U.S. border stations and marauders like Dave Mintz, hired a new president. It was Dave Mintz.

One of the first things Mintz did was bring in American news consultants to design a news format for Global based on surveys and formulas, a method of boosting ratings and revenue which had proven itself in the local U.S. commercial-television wars, whatever

it did to the substance or reflectiveness of the news. (Bill Cunn-
ingham, the vice-president of news who earlier had made such
pretentious statements about Global's news philosophy, quit. Peter
Desbarats, the political commentator, was fired.) Mintz also
appealed to the CRTC to loosen Canada-U.S. coproduction guidelines so that a program could qualify as Canadian content even
though only a quarter of the capital was Canadian. Global, it
seemed, was a big KVOS with a Canadian-content requirement.
Pooling arrangements for a cross-Canada series were made with
western stations, among them CKVU which a few years earlier had
so bitterly criticized "pirate" KVOS and Mintz for their self-serving
mercantile rationalizations.

The Global appointment of Mintz was absolutely inspired. Who
had thought of it? Before, Mintz was president of an American
station in the U.S. beamed at Canadians. Now he was president of a
quasi-American station in Canada beamed at Canadians. He had
the three necessary qualifications: He knew about ratings, he knew
the American television industry and he knew how to exploit the
Canadian market. He could outChercover Chercover.

Global, which was supposed to help defend the cultural border,
had ended up moving the cultural border northward, with itself on
the southern side. Mintz, feeling at home, applied for Canadian
citizenship.

13

The Final Solution
CANADIAN-CONTENT HEARINGS

WHAT DOES A CAPTIVE agency do when — despite its pleading, whining, scolding, coddling, moralizing, attempts at intimidation, and invention of futile manoeuvres like new promises of performance — it cannot get the Canadian branch of the American television industry into the spirit of Canadian content? What happens when this goes on for so long that it becomes too embarrassing to be dealt with by mere renewal hearings? Answer: It holds another Canadian-content hearing.

The capacity of a captive agency to invent new futile exercises as a response to the futility of old futile exercises is unlimited. Canadian-content hearings, in all their vacuous splendour, provide an open-ended, recyclable diversion whose major effect is to keep the agency, and anybody else who takes the hearing seriously, running around in circles for another few years. Such hearings and commissions and associated activities comprise a permanent, cyclical industry in Canada, like pulp and paper and base-metal mining. The original Canadian-content hearing, the Aird Commission of 1929, recommended publicly financed broadcasting for Canada. They got it right the first time. All the rest has been banality.

So the CRTC, after its last few unsatisfying family sessions with CTV and Global and other licensees, announced an elaborate "Canadian Content Review." The announcement, at the end of 1979, listed seventeen different questions, which lent themselves to thirty-four or more possible answers, to "assist parties in making their submission." There followed a long-drawn-out process over more than two years. The deadline for submissions was extended several times.

Eventually 187 submissions were received. Then the commission, during its travels across the country, held public meetings on the issue, trying to beef up public participation. After that came a full-scale formal CRTC hearing in Hull, with verbal presentations by selected intervenors. But unlike the 1970 Canadian-content hearing, there was not much excitement.

Starweek columnist Bob Blackburn followed the proceedings. Blackburn only a few months earlier had been encouraging readers to send in submissions. In 1970 he had been writing for the *Broadcaster,* telling commercial licensees why they should accommodate the Canadian-content regulations then being advanced by a "tough but not unreasonable" CRTC. "There will be conditions attached to licences you've never dreamed of," he told the licensees, and they had better stop living in never-never land and start getting used to it. Now, in December 1981, the coin finally dropped. "The CRTC hearing on Canadian TV content was nothing but a sham," punched the headline.

> Those guys at the CRTC are at it again — worrying their poor little heads about Canadian content on Canadian television. . . . Would it be impertinent today, at the end of 1981, to suggest that enough is enough? Come off it, you guys. You've been fooling around with this for 30 years, at our expense, and you admit that nothing anyone has come up with works. Isn't there something better you could be doing with your time and our money than sitting around trying to square the circle. . . .
>
> If the government thinks the people should get a certain kind of top-drawer programming, then the government is going to have to give it to them. Private broadcasters haven't and won't. . . .
>
> This, of course, would take quite a bit more money than the CBC has, even if the CBC did spend its available funds as efficiently as it ought to. In that fact lies the stupefying paradox of the whole situation. That hearing was held because the government thinks we should have an adequate supply of such programming, yet it refuses to take the one realistic step that would make that supply available. You can say what you will about a few fat-cat broadcasters who are raking it in, or about the fact that private broadcasters won't even *try,* but the fact remains that there isn't enough money in broadcasting in this country to pay for more than a tiny fragment of the kind of TV service we're talking about.
>
> If the government really does want such service to be available, then it must provide it. And if it is not willing to do that, it should not encour-

age its own creature, the CRTC, to indulge in silly shams such as the hearing in question.

It took a mere twelve years to realize what was happening and to recommend a direct line of action, which isn't bad for a critic in one of the world's great centres of derivative thought. At least Blackburn eventually got around to it. Meanwhile, the Canadian-content-review industry lurched on. All that was left by now of the bold nationalist rhetoric was the dregs.

14

The Resistance to Television Advertising

IN THE SPRING OF 1978 the CRTC released a detailed fifty-five-page study entitled *Attitudes of Canadians Toward Advertising on Television*, commissioned by the research department and based on wider research conducted by Market Facts of Canada in 1975 and 1976. It was a blockbuster, so potent in its findings that a couple of years had gone by before the survey results had made their way past internal roadblocks into the light of day. The elaborate data revealed that despite the brainwashing, browbeating, myth-making and mercantile power of American commercial television and its hangers-on in Canada, Canadians still showed an inner resistance to the assault of television advertising on their common sense. One might even dare to call it an independence of mind.

Fully 72 per cent of the survey sample thought there should be at least one channel on which there was no advertising, as against only 13 per cent who disagreed. Fifty per cent said they would rather pay money to watch good programs than to put up with all the advertising on television, as against 34 per cent who disagreed. The most interesting result: 36 per cent agreed there should be no advertising at all on television. Those who disagreed were 48 per cent, not many more than the former. But that was not all. More than four out of five of all respondents considered commercials necessary because it paid for the programs on television. If these people had been aware of the tremendous overheads they had to pay for financing television through commercials, and of the efficiency of doing it noncommercially instead, and also of the different noncommercial financing possibilities, the percentage who favoured no advertising at all on television might well have shot up. The study also found that "a vast majority of Canadians object strongly to the number, frequency, form, content, and manner of presentation of commercials and claim to employ avoidance tactics when commercials come on the air."

The private broadcasting lobby was not pleased with the findings or, more accurately, with the fact that the public should have such an incriminating study available to it.

The data-collection procedures described in the study had been rigorous, involving a painstaking development of the questionnaire, time-consuming personal interviews with respondents, the reading by the respondents themselves of the "attitudinal statements" to which they were responding, and test validation by field supervisors of each interviewer's work. The chairman of the Canadian Association of Broadcasters, only too aware of the implications of the study, summarily dismissed the report as having "no scientific validity." An editorial in the *Broadcaster* snipped away petulantly at the edges, describing the study as a "mouse" and an unproductive bore.

The broadcasters didn't have to worry. The CRTC and the government were not going to do anything about the study. The government instead proceeded to freeze the CBC's budget (a cutback in real dollars), while commercial television was gobbling up more and more Canadian dollars as well as spending ever-increasing amounts across the border for U.S. purchases. There was no CRTC hearing to discuss the implications of the study. The minister of communications did not make a bold speech about it in the usual manner. The ruffled feathers settled down. The study was never heard of again.

15

The CBC, Part II

BUNGLING THE
SECOND CHANNEL IDEA

WHILE THE WHOLE unravelling sequence of television politicking and licensing was taking place from 1968, something else was happening, or rather something else was not happening. Where was the CBC?

The CBC — the "national broadcasting service," the legacy of the Canadian Radio League — was the one body that had the institutionalized muscle to defend the public-broadcasting Canadian tradition. All it could do, however, was flounder around, wasting its muscle. President Al Johnson, who was deeply committed to the idea of the CBC, made earnest speeches about Canadian objectives and the need for more money — and then seemed to make the same speech over and over again until his term ran out. He had the right thoughts but — with one bungled exception, his fight for the establishment of CBC-2 — he had the wrong strategy, or no strategy at all. His speeches merely moved the swamp mud around a little bit.

By the late 1970s liberal intellectual nationalist opinion in southern Ontario — *Saturday Night*, the Association of Canadian Television and Radio Artists (ACTRA), the Council of Canadian Filmmakers (CCFM) — had finally woken up to the possibility that the CRTC was part of the problem, not part of the solution. ACTRA, for the 1978 CBC renewal hearing, for example, wrote that discussion of the CBC alone was useless. The real debate was about the whole system and, by implication, about what the CRTC was doing to it. The CCFM called for a "radical restructuring of the system." The first quarter of its lengthy submission was an analysis of the broadcasting environment.

Johnson made the same analysis. He told the commission that the licensing of additional commercial stations, which for economic reasons had brought in more American programs and given them the best possible scheduling, had made "a mockery of the Broadcasting Act." This statement was not in the least exceptional. That he was making it, was. Neither the government nor the CRTC had had the gumption or honesty to do so.

But watching Al Johnson trying to expand the place of the CBC and public broadcasting in the licence structure — his trying to take the offensive — was an exercise in agony.

The key to the Johnson strategy was the establishment of second CBC television channels in each language. Johnson put forward the proposal in an eighty-page published statement called *Touchstone for the CBC*. Johnson recognized that standing up to be counted was the only way to make progress. *Touchstone for the CBC* was his personal standing up, his manifesto. The document was, as he saw it, "a centrepiece, a philosophical touchstone." It was "a confession, an accusation, and a plan," based on historical reference and practical analysis, including a critique of government drift and of the CBC's own shortcomings. Johnson ended the statement by quoting George Bernard Shaw: "Some men see things as they are, and ask, 'Why?' I dream of things that never were, and ask, 'Why not?' "

The TV-2 proposal, just one of many in the document, was basically simple. The service would be fed by satellite into cable, thereby bypassing the expensive capital requirement of a whole new transmission system. It would for a start be available to a potential 80 per cent of Canadians through cable subscription, and later, through other redistribution, to most of the remainder. It would give the public additional opportunities to see the programming already produced by the CBC, initially distributed on the main network, and would offer some new programming to be phased in over a period of years, the whole at relatively low cost.

But how much exactly would it cost? Where would the money come from? Criticism flared up that the proposal was half-baked, that it had not been sufficiently worked out. These critics missed the point of this kind of statement, which was to put the TV-2 proposal into the world, to get people talking about it and accustomed to it — to float the idea. This is what the cable lobby was doing with pay-TV, despite the fact that very few people were interested in it and that, when it came to Canadian objectives, it was a hoax.

The next step was to get TV-2 licensed and on the air, even if the service might be underfinanced at the beginning. This is how commercial television had established its beachhead and then occupied the territory, in their case to the detriment of the system.

An alert and fighting CBC would have made and fought for the TV-2 proposal, win or lose, in 1957 at the time of the Fowler Commission hearings. Johnson's declaration was only twenty years late. Nevertheless, his effort boldly launching the TV-2 plans and starting the machinery to apply for the licence was no mean personal act and no small gesture.

The main obstacle to TV-2 was the financing. The dollars themselves were no problem for a country the size and prosperity of Canada, and the original budget was set at only $30 million for both English and French channels, a small fraction of existing television expenditures. The problem was the idea of more taxpayers' money being spent, particularly when the first CBC network needed more money. If ever there was a time to make the general argument for public, noncommercial financing, this was it. The argument, moreover, was waiting to be made: exceptional efficiency (1.3 per cent overhead cost), the enormous waste of the commercial financing apparatus (almost 40 per cent overhead cost), the objectionable commercials and commercial interruptions (the CRTC's survey, providing heavyweight ammunition, had become available), the social and cultural reasons for getting rid of commercials. The argument could also have been made in broad, populist terms. But Johnson never made it. He was so trapped by the defence of the CBC's existing commercial revenue that his mind did not get around to it.

One of the most effective ways of generating support for CBC-2 was also lost on Johnson. The Association for Public Broadcasting in British Columbia (APBBC), since its inception in 1972, had patiently been pointing out the logic of locating the headquarters of any new English-language network in western Canada. A western headquarters of a new network was also a matter of deep symbolic and historical importance in a federation like Canada. Although Johnson admitted that the idea was "really a good one," that was as far as it got before he started back-pedalling. English-language headquarters in Toronto had conceived of CBC-2 as part of their extended domain. Johnson had neither the wit nor the inclination to stand up for a western headquarters for the new network. The CBC, not least because of its Toronto-Ottawa parochialism, bungled the job and ended up with nothing at all.

The western headquarters idea would, for a start, have generated support for CBC-2 among some western members of Parliament, and softened or deflected antipathy to the CBC by others. The CBC could have also secured support from at least two western governments at the time (Alberta and Saskatchewan), not discounting other possible provincial support. By wrapping CBC-2 in the flag of equality for western Canada, the CBC would have made any opposition to the idea look like an insult to the West, which it would have been. Westerners would have been hard-pressed to oppose the idea because it recognized the West. Ontario and Quebec politicians would have been hard-pressed to oppose it, particularly since the West had been so long discriminated against and since western power was on the rise. They would not have wanted to be seen as anti-West.

This is fairly basic strategic thinking for a country like Canada. But in order to think it, you have to be thinking in terms of using western power. The group dynamic of Toronto-based CBC-network control functions in terms of resisting western power. Johnson and his surrounding command could not see the best political lever for CBC-2 that they had at their disposal.

Once the CBC-2 prospectus was developed, a request was made to the Cabinet for the additional funding. According to Richard Gwynn, Jean Chrétien was the Cabinet minister who led the way in putting the kibosh on the request. Chrétien, Gwynn said, and other Quebec Cabinet ministers in Ottawa, were still steaming about what they saw as separatist bias on Radio-Canada, and they were going to teach the CBC a lesson. If true, where was the western argument to check Chrétien and other Cabinet opposition? It had been kept out of the CBC-2 plan.

Since the designated funding for the network was not available, the CRTC then turned down the CBC-2 licence application. The commission could have said something bold about the government's refusal to provide funding, to take a small step towards re-Canadianizing the Americanized system. But the hopeless and largely Liberal-appointed commission did not do such things.

"We can't let TV-2 die," a frustrated Al Johnson said after the Cabinet turned down the funding request. TV-2 died.

Johnson's term as president ended the next year, 1982, and was not renewed. The poor guy's big problem was that he kept referring to the Broadcasting Act — and, worse still, believed in it. He was asking for disappointment. And he got it.

Part IV

The Cable Scam

1

The Saskatchewan Case

BROADCAST LICENSING is only half the television story and, despite its hypocrisy and futility, not the more outlandish half. The cable take-over scam is.

It began in Saskatchewan in the early 1970s. The government of Allan Blakeney announced that cable television would be created as a crown corporation and would take advantage of facilities owned by Saskatchewan Telecommunications (Sasktel), the provincially owned telephone system. Nonprofit community organizations would run the local programming service after obtaining a licence from the CRTC. The announcement was made in late 1972 by John E. Brockelbank, minister of government services and telephones. Cabling had already begun in Regina in anticipation of cable television. Shortly after, the idea of a crown corporation separate from Sasktel was dropped.

Saskatchewan took the view, as Blakeney was later to explain, that federal agencies did not have clear constitutional jurisdiction over cable television, as they did over radio and television stations. "Cable television is simply not an interprovincial function," he said, and Saskatchewan should have a major role in determining some of the programming carried on its cable systems, particularly programming of an educational nature. The community channel, taken separately as a closed-circuit service, also fell under provincial jurisdiction.

Saskatchewan's position, however, was far more than just a theoretical jurisdictional one. The province was entering into cable television, through Sasktel, to ensure that excess profits would not be taken out of larger centres like Regina and Saskatoon and that service would be extended to smaller centres as quickly as possible through cost-averaging. "The best service at the lowest cost for the largest number of people" was how the government described its policy.

This concept of a nonprofit, service-at-cost structure extending communications to all parts of the province at equal subscriber cost as quickly as possible had deep historical roots in Saskatchewan. Public ownership of the telephone system there, as in Alberta and Manitoba, went back to the early years of the century. Sasktel was an expression of the Saskatchewan character. The Saskatchewan government also knew that the province could not count on federal regulation to serve the public interest. Some of the existing cable systems were notorious for the monopoly profits they were ringing up, thanks to an obliging CRTC. There was little chance of a political remedy from a tangled federal scene, particularly with its loose consensus-style government, extremely vulnerable to vested-interest pressure. Saskatchewan politics, clearer and closer to home than Ottawa's, polarized by the presence of a major social democratic party and characterized by a relatively high degree of public awareness, provided much better long-range protection for the public interest.

There were other, technological, reasons for Sasktel cable ownership. Advances like fibre optics opened the possibility of telephone, telecommunications and cable services being integrated into the same local distribution system.

In southern Ontario cable was already established. At the very least, the amplifiers along the lines and the drops into houses and apartments were owned by the cable companies. This cable ownership system was CRTC central Canadian dogma, but it meant nothing in Saskatchewan where cable, with minor exceptions, was not yet established and where the telephone company was publicly owned. Saskatchewan regarded as extraordinary folly the idea of separate and parallel facilities when costs could be reduced by integrating services. If Ontario and Quebec wanted to pay for duplicate distribution systems, it was their loss; nonsensical duplication was not going to be imposed on Saskatchewan.

The province insisted that the CRTC lacked the authority to make decisions that would influence the future of Saskatchewan telecommunications, as distinct from broadcasting, and to undermine the province's long tradition, supported by all political parties. In this view, future closed-circuit uses of cable, like security alarm systems and teleshopping, fell under provincial jurisdiction.

The CRTC's position to that time was that cable licensees needed to own the amplifiers, the drops to houses and the head-end receiving the signals, in order to ensure that they were able to deliver the required television channels free of interference by, say, a province or a telephone company. The Saskatchewan government was prepared to offer safeguards to Ottawa and to the CRTC assuring priority carriage of the programming channels of CRTC licensees. The provincial government also gave assurance to Ottawa that no action would be taken "to influence or jeopardize the autonomous role under CRTC regulation of those licensed to create and provide the content of the [cable television service]." There would be clear separation between "carriage" (Sasktel's distribution system, available on a common-carrier basis) and "content" (in this case the programming provided by the CRTC's licensees). Only the Saskatchewan arrangement accommodated both the traditional telecommunications policy of the province and the federal Broadcasting Act.

The provincial government, a few months after its original announcement, also set guidelines for the ownership and control of cable television operations in the province, including: (1) nonprofit ownership and control by subscribers of their cable organization through share ownership or other representative membership arrangements; (2) free and reasonable access to program production and distribution facilities; (3) programming advisory councils to promote the participation of the community in program production, and (4) wide community support. Co-operation among cable television organizations in different communities, to share resources and help smaller organizations, was also envisioned.

One of the major advantages of Sasktel's building and operation of the distribution system was that the cable licensees themselves would not be required to raise the relatively large amounts of capital for the distribution plant. This opened the door for community-based organizations to apply for licences, and for the CRTC to consider them as trustees for the licences without being bogged down by a concern with their ability to raise and service capital. The Saskatchewan guidelines were not brought forward simply to ensure that cable organizations should be nonprofit. They were a beginning attempt, through the small jurisdictional leverage that Saskatchewan had, to make ownership and control of media more representative and to do something about the way federal broadcasting policy, or the lack of it, had casually discriminated against Saskatchewan in the past.

The cable co-operative organizational framework was part of the established co-operative mainstream tradition in Saskatchewan, which like public telephones also went back to the early years of the province. Indirectly, through the powerful credit union movement, the provincial Department of Co-operation and the Co-operative Guarantee Act, the framework was part of a large economic and financial infrastructure, including the government's bond rating.

The first group of cable licences to be allocated were for Regina, Saskatoon, Moose Jaw and North Battleford, using a distant head-end and microwave relay for U.S. signals. Weyburn and Estevan, near the border and directly within reach of the signals, already had cable systems. Cable co-operatives were organized by interested individuals and group representatives in each of the four new cities to apply for the licences. The best, one might say the greatest, of these co-operative applicants was the Saskatoon Cable Co-operative, though the others were exceptional in their own right.

In early 1974 the embryonic Saskatoon co-operative opened an office and hired an executive director with $500 contributions by group members and a $5,000 start-up grant from the provincial government. The president was George Dyck, also president of the Saskatoon Credit Union and a city alderman. The executive director was Courtney Milne. A committee structure was set up: technical, program development, organization and finance, and information. The first two committees drew for their leadership on people at the University of Saskatchewan in the extension department and in educational communications. The program development committee quickly grew in size to thirty members and held frequent meetings. By the early summer of 1974 extensive and detailed programs had been developed for almost any size of program budget. An organizational structure evolved whereby half the directorships were to be allocated to group members, from whom the strongest demand, and potential, for programming would come, and the other half to individual members. The information committee began producing information leaflets and brochures. Workshops were held. A tent display was mounted at the Saskatoon Industrial Exhibition, and another display was organized the following year. A Media Access Training Program offering ten different production courses was established through the Saskatoon Region Community College. The training program was based on an extensive programming questionnaire distributed to member organizations.

On a visit to CRTC offices in Ottawa, Saskatoon Cable Co-operative representatives were told by staff members that one of the most important qualifications for getting a licence was community support, which had in fact already been developed as an integral part of building the organization. The original organizational meeting, many months before incorporation, had approximately one hundred group representatives. By the time of the licence hearing in February 1976 Saskatoon Cable Co-operative had the endorsement of 145 community organizations, as well as seven hundred individual members picked up in the last phase. In every case, the endorsing organizations also joined the co-operative. Each organization was also committed, by membership, to elect a delegate to represent them at the co-operative's meetings.

Parallel to this organizational effort was the development of a Program Advisory Council to allow local groups as much control and influence as possible in planning the community channel programming, to provide a forum for each "programming area," to meet and exchange views and ideas and to divide the available funds among the different program areas in the channel's operation. The council was made up of representatives of nine individual program councils (community service, religion, human rights, multicultural, education, arts, sports and recreation, labour, and special programs), each with its own chairperson and representatives of participating member organizations. Access to the community channel would not be limited to members, but would be open to any other group or individual through the program manager.

A random telephone survey done by the co-operative after this process of public involvement and education found that 73 per cent of the community knew about the organization and 88 per cent were in favour of the concept of nonprofit community control. The formal licence application filed with the commission in late 1975 was a fully documented, five-hundred-page bound volume, most of it relevant. Saskatoon Cable Co-operative was the best cable licence applicant the CRTC had ever had.

When it came to the general aims of the Broadcasting Act and its Canadian objectives, Saskatchewan also presented the CRTC with a rare and fortuitous situation. The Saskatchewan government was the most nationalist, in the Canadian sense, in the country. Allan Blakeney was probably the only head of government at the time who understood the importance of Canadian objectives in broadcasting and filmmaking, who realized at the same time that the key factors undermining these objectives were structural economic ones. The Saskatchewan government, in its own extensive submission to the CRTC on the cable licensing, openly and explicitly defended the proposition that broadcasting and cable should be seen first of all as instruments of public purpose rather than commercial profit. This was the traditional ideology of Canadian broadcasting policy — Saskatchewan cited the Aird Commission and the Fowler Commission — and if wishy-washy colonial Ottawa had largely defaulted on it, well, here at least was official provincial support for applying the policy.

The Saskatchewan idea of the separation of carriage and content in cable, however, was a situation devoutly undesired by the private, nonco-operative applicants on the scene, of which there were several for each city. Take away the millions of dollars of assets in physical plant, which the licensee would have got the subscribers to pay for anyway, and there goes the risk and investment excuse for the licensee to be a profiteer. But what else was the point of applying? "Why are you in cable?" one of the private applicants asked George Dyck, president of Saskatoon Cable Co-operative. "What are you going to get out of it? I'm in it for the money, myself."

Nor was the cable lobby in general enamoured of the Saskatchewan prospect. Once the mythology of the heroic and especially deserving cable-system entrepreneur went down in Saskatchewan — because Sasktel would be building and operating the distribution system — what was to keep the self-serving mythology and its high rate of return propped up elsewhere? The lobby attacked the Saskatchewan plan for Sasktel hardware ownership in terrorizing fashion to strike fear into the hearts of the federal guardians.

When the licence hearing eventually took place in February 1976 in Regina, Jerry Grafstein, Toronto communications lawyer, appearing for both the Canadian Cable Television Association (CCTA) and several private applicants, diligently raised the bogey of government ownership in general and then provincial government control of the cable licensees and, by implication, God knows what other fearful prospects. "I'll select the one group that applies," he sardonically mimicked the Saskatchewan government, "one co-op for each area, and I'll set up eight guidelines, but please, don't believe that I have any desire to exercise my control." That the guidelines would have guaranteed wide-open, representative community ownership of cable licensees rather than control by a handful of profiteering private board members, like Grafstein's clients — and this was the point of the guidelines — he did not deign to mention. (Grafstein, now Senator Grafstein, was also a City-TV and CKO shareholder, and Liberal operative.) The CCTA raised the spectre of "a provincial government takeover of any service — newspapers, radio, television stations, for example — which might not be economically viable in every small community." Israel Switzer, a Toronto engineer in the service of Agra Industries (the Weyburn and Estevan licensee, and a Saskatoon applicant) commented of Sasktel that "we can see a future in which touch telephones will be available only to card-carrying NDP members, and coloured telephones will be the exclusive prerogative of co-op members."

A hearing like this exists so that Toronto lawyers and promoters and their openly mercenary clients and their Ottawa lobbyists can go to Regina and lecture the people of Saskatchewan about democracy.

Switzer, in a later intervention to the CRTC, with his client's back against the wall, was more to the point. "Underneath all of this, all these arguments, is really a matter of money and power. What is at stake . . . is a dispute over the right to invest capital . . . because along with the investment of capital comes the expectation of profits. In addition, there's a consideration of power. Money and power." The proper beneficiary of publicly allocated money and power, in the Agra Industries and cable lobby scheme of things, was not the public, but a few special interests that were playing the game — and the CRTC should look to it.

The other effect of separating carriage from content would be a change in the character of cable licensees, since they would no longer own and operate the distribution system. They would be marketers and bill collectors. They would also own and operate the head-end and switching equipment. These were routine functions. What was left was the community channel and other original Canadian programming, which together would be the major defining function of the cable licensees, making their internal orientation local and Canadian instead of American. In this scheme of things, further, a profiteering private company was out of place. The non-commercial programming function is a money-spending function looking to ways of expanding program production, not a money-making one.

This was foreshadowed in the development of the co-operative applicants in the four cities. The co-operative, for example, had committed all surplus revenue to program production. Plans for a program production trust under a newly formed Saskatchewan Co-operative Cable Federation also took shape, for programming over and above the community demands. As originally conceived, the trust involved an allocation of only 3 per cent of gross revenue, but the idea was of crucial importance: the allocation of cable revenue to professional Canadian program production for general distribution. The program production trust, at the same time, was a direct step to help Saskatchewan participate in Canadian program production

and to do so in public broadcasting terms, where existing broadcasting structures had neglected and discriminated against Saskatchewan participation. For these reasons, special historic importance was attached to cable and the co-operative framework in
Saskatchewan. The participants looked to cable in pioneering
ways that went well beyond the token community channel elsewhere.

This Saskatchewan initiative — with nonprofit structures, maximum allocation to programming, using a portion of cable revenue
for professional program production (thereby opening the cable
financing mechanism to Canadian objectives), and direct ownership
and control of a media sector by the citizenry — constituted the
most important initiative in Canadian broadcasting history since the
Canadian Radio Broadcasting Act of 1932.

The cable co-operatives' structural and organizational advantages
were great, and the particulars of their applications far outdis-
tanced the competition. This was obvious at the licence hearings,
too.

The CRTC decision came out in the summer of 1976. It rejected
the traditional Sasktel service-at-cost, cost-sharing, crown-owner-
ship plan for the distribution system. Moreover, although it
awarded licences to the co-operatives in Regina and North Battleford, it denied them to the co-operatives in Saskatoon and Moose
Jaw. In the Saskatoon case the decision rejected the best and strongest application the commission had ever received. It also destroyed
the idea of a federation of nonprofit co-operative licensees on a
province-wide base, which could generate the political impetus and
the financial wherewithal to undertake new program production
initiatives. The decision closed the door on a badly needed new
Canadian direction and cut down at the roots the creation in at least
one region of a media sector owned and controlled directly by its
citizen-users, isolating subscriber ownership in just two centres —
one of them, North Battleford, a small one at that.

The Saskatoon licence went to a syndicate headed by an ex-Saska-
tonian who parachuted in from Bramalea and the Ontario cable
industry. The syndicate's application was pro forma. They had no
endorsements of public support. Their projected average programming expenditure over the first five years was $117,000 annually,
compared with $268,000 annually for the co-operative. Other com-
parisons were similar. The particular differences in the Moose Jaw
case between the co-operative and the designated licensee were not
in all respects quite so dramatic, but were still substantial.

Different layers of Ottawa territorial presumptions, arrogance, ignorance and bureaucratic *amour-propre* lay like sedimentary deposits underneath the notorious Saskatchewan decision. Then there was the question of co-operatives. It was one thing for the bureaucratic expedition from southern Ontario to acknowledge politely the co-operative movement in Saskatchewan, in the way that metropolitan tourists would acknowledge and praise folklore in the hinterland. But when it came to cable, the co-operative tradition in Saskatchewan was beyond even Harry Boyle, who had had an early connection with co-operatives in Nova Scotia. "When you take a co-op which does not have an economic thrust to it, which has, in effect, a kind of social thrust, how do you manage to keep people interested?" asked Boyle. It seemingly did not occur to him that "social thrusts" interested people, co-operatively minded people most of all. How had the cable co-operatives managed to become so effectively organized in the first place? Besides, in addition to the co-operatives' other advantages, what other structure would create an internal lobby with voting power — the groups and individuals doing programming — whose interest was in spending more money on programming? Did the Broadcasting Act not have a "social thrust"?

There was also a hidden factor: the existence of Quebec, or, more precisely, the paranoia in Ottawa about Quebec. Quebec had been conducting a guerrilla war against Ottawa for jurisdiction of cable. The spectre of Quebec controlling any part of communications and blocking the implementation of Canadian policy filled Ottawa heads. Separatism was on the rise and plagued the very thought processes in the federal capital. The whole town was battening down the hatches against leakage of federal jurisdiction, in order to save the unity of Canada, it was purported. In cable, the Maginot line was ownership of the amplifiers and drops by the CRTC's licensees, under federal control.

The Saskatchewan proposal for Sasktel hardware ownership entered into this addled environment. It was taken not as coming from one of the most pan-Canadian of provinces, Saskatchewan, but as if it came from a little Quebec, somewhere out there on the prairies, which would use ownership of the cable carrier for God knows what quasi-separatist, balkanizing purpose. It could even block cable distribution of the CBC, the national system.

Although all the co-operative applicants were prepared to meet the requirements of the CRTC's hardware ownership policy if necessary, and said so when questioned at the hearing, only the Regina Cable Co-operative went out of its way to accommodate the option as a possibility. So the CRTC gave it the first co-operative licence in the politically calculated split. The North Battleford licence was the smallest, the farthest out of the way and the least important. The co-operative applicant there, moreover, was the only one that had applied for North Battleford by itself. So it got the licence. The Saskatoon and Moose Jaw co-operatives, exceptional as their applications were, got what was left over — the short end of the stick.

On the heels of the licensing decision, the government of Saskatchewan and the Saskatoon and Moose Jaw co-operatives filed petitions with the federal Cabinet, asking it to set the decision aside. Ottawa, over the signature of the prime minister, replied with the same smug and remote reasoning as the smug and remote reasoning of the CRTC decision. Saskatchewan and the co-operatives also began court action challenging Ottawa's jurisdiction. But Saskatchewan had an ace in the hole: It denied the use of Saskatchewan's streets and lanes to the CRTC's licensees, without which they could not string cable and begin operating. Sasktel's own cable construction program picked up speed.

Saskatchewan had clearly warned the CRTC at the licence hearing that under no circumstances would CRTC licensees be given access to the streets and lanes for their own cable. Toronto lawyer Grafstein, speaking for his clients, encouraged the CRTC, however. "Yesterday, I recognized a distinct softening in the position of Sasktel," he said. "I think . . . that an accommodation will be reached if the commission upholds its policies." Perhaps, for the CRTC, a Toronto Liberal lawyer was a greater expert on Saskatchewan than the Saskatchewan government. The other private applicants also encouraged the CRTC. Once the dust had settled, the Saskatchewan government, being made up of reasonable people, as these applicants put it, would negotiate a satisfactory agreement (that is to say, the Saskatchewan government would capitulate). It was a pleasant scenario, and one that came naturally to a lobby that assumed, one way or the other, that it should get its own way.

Saskatchewan then actually did what it said it was going to do. In effect, it told Ottawa to go to hell. There was not too much Ottawa could do about it, not only because of provincial jurisdiction over streets and lanes but also, and maybe more importantly, because of the politics of it. The Saskatchewan Conservatives, the official opposition, also favoured the provincial position. The provincial Liberals made the anti-Saskatchewan argument at their loss. Saskatchewan locked the door to the CRTC's licensees for a whole year, until Ottawa finally came to its senses.

There was another lever working against Ottawa's presumptions. The members of the Saskatoon Cable Co-operative met, following the CRTC decision, to decide whether simply to accept what Ottawa had done to it and fold, or to undertake a counter-strategy. They decided to proceed with a closed-circuit pay-television operation transmitted over cable, independent of the CRTC, since closed-circuit television that did not take signals off-air fell under provincial jurisdiction. Saskatoon was joined in discussions by the cable co-operatives in Moose Jaw and North Battleford. The Co-operative Programming Network, as it came to be called, was given political support and seed money, and later a substantial financial guarantee, by the provincial government. The original idea of the service was four or five channels: conventional, children's, educational, community, and a movie channel for a supplementary rate. The economic core of the package would be entertainment programming, largely American, for which rights would be bought. Available Canadian programming would also be used. Revenue generated by the service would be applied to the community channel and to a program production trust originally proposed to the CRTC. The service would be in addition to the conventional cable services when they came on stream.

Ottawa was distressed. Minister of Communications Jeanne Sauvé, in Churchillian flight, declared that Ottawa would fight the Saskatchewan plan to the finish. She was confident that Ottawa had the legislative basis to win the battle. She was also confident Ottawa had public support on its side. (She did not have the support in Saskatchewan to fill a pail with Saskatoon berries, but what the heck.)

The Saskatchewan plan, she said in an interview to the press, could kill the orderly introduction of pay-television across Canada, particularly if other provinces chose to follow Saskatchewan's lead. The disastrous and far-reaching consequence of the plan, said Sauvé, would be to siphon off millions of dollars that would have been turned back into expanding the production of Canadian television programs. The Toronto *Star* took up the cry: "National TV policy must cover cable."

The Saskatchewan minister responsible, Consumer Affairs Minister Ned Shillington, obligingly volunteered to co-operate with Ottawa and to discuss a pooling arrangement for the profits, but the revenue generated from Saskatchewan pay-TV was not just going to go to subsidize "eastern broadcasters" and underwrite production in Toronto in the guise of a so-called national policy.

"There has to be some recognition of Saskatchewan's needs," said Shillington. "Co-operation is a two-way street. It always seems to be a one-way street with Ottawa."

Whoever lifted Sauvé onto her high horse had forgotten to tell her, or had not understood, that the CRTC had in Saskatchewan just destroyed a real initiative for Canadian program production and had brought these new troubles onto Sauvé's head. Instead of a safe-lock on closed-circuit pay-TV in Saskatchewan, and nationalist Saskatchewan as an ally in the fight then going on against Quebec and Ontario over pay-TV jurisdiction, Ottawa ended up with a belligerent Saskatchewan in no mood to surrender closed-circuit jurisdiction of any kind. Saskatchewan had had one lesson too many in Ottawa betrayal and incompetence. Now the province was seeking out Quebec, Ontario and other provinces to form a common front against Ottawa.

The closed-circuit pay-TV plan also undercut public pressure in Saskatchewan for the immediate introduction of conventional cable television and put private CRTC licensees on the spot. They were faced with the possibility of the closed-circuit service establishing itself before they even got a foot in the door.

In November 1976, only two months after rejecting Saskatchewan's petition regarding the CRTC decision, the federal government, its nervous system stimulated by the fiasco it now faced in Saskatchewan, rejected a similar CRTC decision for new Manitoba licences and signed an agreement with the Manitoba government that allowed the crown to own the cable distribution system. Manitoba, like Saskatchewan, had owned its telephone system since the turn of the century and made the same defence for protecting its traditional role. As quick as a wink, the agreement ended a three-year struggle by Manitoba. In conceding the point, the federal government retained jurisdiction over programming services. A ruffled CRTC and its chairman Harry Boyle huffed and puffed that the agreement "would erode our independence if we simply accepted it." But a major policy matter like this should never have been decided by an appointed regulatory agency in the first place. The powers of Cabinet to set aside a CRTC decision existed exactly for this kind of situation.

"Carried to its logical conclusion," Boyle then offered, "a cable operator would become merely a bill collector with no responsibility to the community." This was the old CRTC line. Carried to its logical conclusion, if the only reason for the existence of cable licensees was to gouge the public and build little empires through ownership of cable hardware, then why, in a situation like Manitoba, bother with cable licensees at all? Why bother with the CRTC at all?

The Manitoba agreement had a magic radiation effect on the private cable syndicates in Saskatchewan, who went through a change of life. The head of the Saskatoon licensee company mutated from a moralistic apologist of the old Ottawa position into a frantic diplomat trying to get Saskatchewan and Ottawa to settle.

Shortly after the agreement, Saskatchewan consumer affairs minister Shillington met with Sauvé. He again offered to plan Saskatchewan pay-TV so it would fit into the national system, but he declined to surrender jurisdiction over it to the federal government as Manitoba had. The meeting was cordial and conciliatory.

"I think if we had had [the meeting] a long time ago, we'd have settled a lot of things," Sauvé said. Here was the final clue about Ottawa's incompetence and ignorance. Sauvé did not mention that Saskatchewan had initiated such meetings, also cordial, in 1973 and 1975, had also made its position clear in its submission to the CRTC and its appeal to the federal Cabinet in 1976; the federal Cabinet in that latter connection could have called a meeting, but instead refused and turned down the province's petition at the last possible moment, as the deadline for Cabinet action passed, so that further discussion would be pointless. Nothing of substance was said at this meeting, either, that had not been said before. Without active and politically secure Saskatchewan resistance, Ottawa would not have settled anything on the issue, had meetings occurred daily and twice on Sundays for the next fifty years.

The Manitoba-Ottawa agreement soon led to the CRTC also giving way on Sasktel ownership of the cable distribution system in Saskatchewan. What happened, though, to the injustice done to the Saskatoon and Moose Jaw co-operatives? Their legitimate claim to the CRTC licences was left hanging in the wind. The Saskatchewan government did not hold out for them, or for a new hearing. The crucial ideas of the province-wide programming federation and the program production trust were also left hanging, since the closed-circuit scheme, onto which the ideas were grafted, was too financially dubious to support them.

In the Manitoba case, where the decision for new licences had been set aside because of the hardware dispute, a new licence hearing was held, open to all applicants. The Westman Media Co-operative, which had not applied the first time, was successful in acquiring the licences for a large rural-urban area in western Manitoba, including Brandon and Dauphin. In the Saskatchewan case, where the change in circumstances corresponded, but where the original CRTC decision had not been set aside by the Cabinet, the commission blocked any reconsideration.

The Saskatoon and Moose Jaw co-operatives pointed out that since their applications had strong local support — one of the commission's principal criteria — and were similar in all other respects to the North Battleford and Regina co-operative applications, their denial must have rested on the question of hardware ownership. The hardware arrangement now agreed to by the CRTC was precisely the one the two co-operatives had recommended, whereas the hardware arrangement that the private applications had been based on had now been ruled out.

The bitter irony of the two leading co-operatives now being on the outside was not lost. By all logic and duty, aside from fairness, the CRTC should have held new licence hearings. There was also a formal practical procedure for it, as the first licensees had not fulfilled the hardware conditions of their licences.

Instead, the CRTC merely amended those licences. It even refused to hold a hearing on the amendment applications, despite requests from the Saskatoon and Moose Jaw co-operatives. The commission was not obliged by law to hold a hearing to amend applications if it decided a hearing was neither necessary nor in the public interest. The two co-operatives could go hang.

The closed-circuit pay-TV system did not have the economic base in its eventual three cities of operation (Regina, Saskatoon and Moose Jaw) to be viable and at the same time to meet its objectives. Conventional cable had first call on the discretionary television dollar in most households. The experimental nature of the closed-circuit operation plagued it with start-up problems — from signal protection to the acquisition of program rights — not of the operation's own making. With its losses piling up, the operation was taken over in 1979 by a consortium of the conventional cable licensees (40 per cent), the big co-operatives (30 per cent) and the Crown Investments Corporation, Saskatchewan's crown holding company (30 per cent). The service was stripped down to a single channel, largely a first-run American movie channel, the most marketable part of the old package. "It's a system whereby Saskatchewan farmers subsidize American movies," one veteran cable co-operative partisan commented acidly. But the viability of the closed-circuit scheme was always doubtful. The scheme from the start was a third- or fourth-best, a last recourse for the shattered co-operative federation and its broken Canadian plans.

Instead of a historic pioneering Canadian initiative, well-organized, rooted in society, politically buttressed and drawing on its own native dynamic, what Canada got in Saskatchewan, thanks to the Ottawa machinery and its cogs like Harry Boyle and Jeanne Sauvé, was an American movie channel. The valuable social energy of Saskatchewan was discarded. Sasktel, the object of Ottawa's futile gesticulations of power, kept the hardware all the time.

2

Capital Cable Co-operative

THE PEOPLE FIGHT BACK

IN THE SPRING and summer of 1975, the year before the Saskatchewan cable hearing, the Association for Public Broadcasting in British Columbia (APBBC) began a guerrilla-style action to bring about the necessary reform of the licence structure. It led to one of the most extraordinary cases of evasion in the history of Canadian public administration.

After two years of dealing with the CRTC, the APBBC had come to the working hypothesis that evasion was the CRTC's business. Since the commission would not consider the general policy proposal for creating a new Canadian programming sector through cable, the next step was to apply for specific licences as they fell due and build the programming sector, by force of example, from British Columbia outward. Several major B.C. licences, including the ones for Vancouver and Victoria, were to expire in March 1976. There was only one problem — the CRTC refused to hear competing applications when the old ones ran out. And as a Vancouver journalist was to comment later, "unless somebody murders his mother," renewal was virtually guaranteed.

So the first purpose of any application organized by the APBBC was to provide the grounds for a test legal action against the commission when the application was returned without a hearing, as was expected. Separate meetings were held in Vancouver and Victoria, where the association had a sufficient concentration of members and organizational support, to see if there were enough people idealistic enough, and angry enough, to take on the CRTC, regardless of the odds. The Vancouver meeting decided it was hopeless. The smaller Victoria meeting decided to go ahead.

The original working nucleus consisted of three Victoria APBBC members. I became organizer, making a fourth, since there was no money to hire anybody with the necessary background. This was how Capital Cable Co-operative began.

One of the three members, Kay Lines, almost single-handedly built up a list of sponsors, many of them well-known names, to get the organization off the ground. The others in the original group were Beth Dickman and Adrian Biffen. The president of Capital Cable Co-operative and one of the founding incorporators was John Young, a controversial, nationally known educator from Campbell River and in 1975 a Victoria businessman. Young had been president of the Campbell River TV Association (CRTV), the country's pioneering subscriber-owned cable system. CRTV, going strong since 1956, had introduced a $1-a-month rate for pensioners; offered additional outlets at no extra monthly charge; had been among the first to establish a fully equipped community programming facility, particularly unusual for small systems (less than five thousand subscribers) in those days; had a high-percentage community programming budget; had built an uneconomic line to a small outlying community to share the service; was fully unionized, and all at a low monthly rate. Young had been instrumental in many of these developments.

The most prominent sponsor of Capital Cable Co-operative, soon to become vice-president, was Peter Pollen, the mayor of Victoria. Pollen was the Ford automobile dealer in Victoria, and before that had been western Canadian business consultant with the Ford Motor Company. He was a true individualist, outspoken in defence of the public interest. His administration was one of the most enlightened in North America.

Pollen's support of Capital Cable was not accidental. In 1972 the City of Victoria, with Pollen as mayor, led the opposition to a rate increase sought by Victoria Cablevision. In 1974 Pollen appeared at a CRTC hearing in Vancouver in connection with the renewal of two Victoria radio stations and the Victoria television station. As well as giving an uninhibited critique of the stations, he criticized the commission itself for its failure to regulate the "strident commercialism" of certain stations and for allowing the "sale of licences to the highest bidder."

One of the reasons the APBBC had turned to the cable subscription mechanism as a means of funding a new noncommercial programming service was because, even by the early 1970s, southwestern British Columbia was heavily cabled. The subscription level in Victoria was the highest in the country and in the world, hitting 92 per cent in 1976. The idea of financing a new service through cable came naturally there, whereas in most other

places in the country the general presence of cable, to the point
where it was considered standard household equipment, was still
several years away. What did not fit in Victoria was that the system, using a monopoly public licence, should be owned outside
Victoria and be used as a milch cow for a Vancouver holding
company, Premier Cablevision.

Premier owned seven cable systems in Canada — in Vancouver,
Victoria, Coquitlam (a Vancouver suburb) and four systems in
Ontario (York Cablevision and Keeble Cable Television in Greater
Toronto, Oakville Cablevision, and Borden Cable Television) — as
well as other holdings, including a majority interest in two systems
in Ireland. Premier's Vancouver system, with 201,000 subscribers,
was one of the largest in the world. But the Victoria operation,
because of its high penetration, the physical proximity of households and the system layout, was also highly lucrative per subscriber. In 1974 the profit before taxes was an amazing $1,020,738
on total income of only $2,563,365. The 1975 profit before taxes
was $975,554 on total income of only $2,810,412. This worked out
to a return on shareholders' equity before taxes of 48 per cent in
1974 and 38 percent in 1975. These were extraordinary rates of
return for a publicly licensed monopoly utility. This shareholders'
equity, in turn, had been built up by retained earnings from previous excessive profits. The original investment in Victoria Cablevision — the total share capital showing on the books — was a nominal $120. Premier also took money out of the Victoria system and
other subsidiaries through intercompany charges. In the 1975
financial year, for example, Premier charged the Victoria system an
"administration" fee of $293,000. The increase alone in that cate-
gory from the previous year, $112,000, or 61 per cent, was more
than the system's entire annual operating expenditure on community programming.

People interested in community programming, many of them
directly involved in the programming, realized that Victoria cable
subscribers, who were pouring money into the system every year,
were creating a tremendous potential, but the available revenue was
being diverted. Subscribers, through their monthly payments, had
paid for the system in the first place.

The city, treated like a satellite of Vancouver, did not have its own
CBC television station. The station it did have, a BCTV-owned CBC
affiliate, was making no cultural contribution. Local subscriber ownership of cable would repatriate all available revenue for Victoria
artistic and community life, new professional television production,

lower rates, neighbourhood access and other possibilities. It would underwrite producers and performers. It would make a cultural and economic contribution that would not be generated any other way, least of all by the decision-makers of the CBC and CTV in Vancouver and Toronto. It would also provide local financial control and full financial disclosure, freeing the system from Premier's intercompany manoeuvres. This desire for local financial control went back several years. For many people it was the most compelling part of the Capital Cable licence proposal.

In the proposal, community programming, exclusive of depreciation on studio equipment, would be allocated 10 per cent of gross revenue, over three times what Victoria Cablevision was spending. This included direct allocation to volunteer community programmers, on a sliding scale according to the nature of the program, to help improve programming quality. Capital Cable also adopted, from the work being done by Saskatoon Cable Co-operative, the idea of a program council with power to decide on production priorities, scheduling, allocation of resources and the use of equipment, under the budget assigned to it by the board and with a direct interest in improving quality. Plans were also drawn up for member-viewers' participation as guest commentators and interviewers, for two-way distribution from neighbourhood centres, for viewer feedback about both the community channel and the cable organization, and for an extensive information network to keep people in touch with what was on the community channel. All money available over and above operating costs and the community channel would go into the new professional program production sector.

Equally important was the social and political innovation — a citizen-owned sector of the media. Every cable subscriber would be a shareholder-member and have full and equal voting rights Through the election of the board of directors, subscribers would have a measure of control over their cable operation. Shareholder-members would own equity in the system; Capital Cable would be buying the system from Premier if it got the licence. This equity, representing the subscriber's share of net assets, would be fully refundable when the subscriber left Victoria or no longer wanted cable.

The Capital Cable application for the cable licence for Victoria, Saanich, Oak Bay, Esquimalt and district was filed by hand in Ottawa on October 31, 1975, all fifteen copies of it. The application was more than a pro forma one. It included an outline of the rationale and the Canadian objectives of the organization and of the advantages of nonprofit subscriber ownership; the proposed incorporation particulars; detailed financial material complete with five-year financial projections; a description of how the changeover of assets would take place; an extensive outline of community programming plans; arrangements for public participation and the formation of the program council; a sample programming schedule and finally a brief discussion of the plans for new professional television production. The purpose of the extensive application was twofold: first, to work out in detail and put down on paper Capital Cable's licence proposal for its own sake, and second, to demonstrate the advantages in enough detail that the CRTC might reconsider its policy of excluding competing applications.

A month later the entire bulky package of fifteen copies came back in the mail, unread, with a covering letter repeating the policy. No explanation for the policy was given. The anticipated legal action followed, aimed at obliging the CRTC to hear the Capital Cable application. Capital Cable was represented by David Lisson, a young Victoria lawyer with his own firm. Working on donated time, he did not manage to get the case before the trial division of the Federal Court of Canada until February 2, 1976, only three days before the scheduled renewal hearing in Victoria. The case was heard in the Federal Court chambers in Vancouver by Mr. Justice Jean-Eudes Dubé, a former Liberal Cabinet minister from New Brunswick.

The court hearing took place the same day that CRTC hearings in Vancouver began on a variety of licence matters, chairman Harry Boyle presiding at the latter. The commission hearings that day were taken up with a stormy session on the licensing of a CBC French-language television station in Vancouver and the possible bumping of an American channel off basic cable capacity. Members of Parliament Simma Holt and John Reynolds, plus a couple of radio talk shows, had unleashed a campaign against the licensing of the station, a campaign full of undercurrents against the French-language, Quebec domination of the federal government, and bilingualism and biculturalism. A coupon mail-in organized by Reynolds generated 28,000 signatures. Radio station CKNW gathered 116,000 names, which were dumped in front of the commission. Although the licensing of the French-language television station also had its defenders, the emotional pitch of the hearing was exceptional. The CRTC, as the incarnation of the federal doctrine, was in the direct line of attack. Holt was particularly vociferous. She brought a small army with her to the hearing and passed out lapel buttons reading "CRTC Go Home." It was the regulatory equivalent of a day under shelling.

Meanwhile, in Federal Court, Lisson was making his argument. It was based on Section 3 of the Broadcasting Act, the policy section, with its ringing clauses on Canadian objectives and its declaration that "radio frequencies are public property." The argument for "natural justice" was also put. There was no case law specifically applicable. Victoria Cablevision — or rather, the parent company, Premier — was allowed by the CRTC to join with it as a party to the action. The symbolism of the CRTC and Premier locked together in court seemed shocking, but it was appropriate. Dubé then retired to consider and write the judgement so it would be ready the following morning.

Dubé found in favour of Capital Cable Co-operative and granted it a writ of mandamus. The writ obliged the CRTC to hear Capital Cable's application before issuing any renewal to Victoria Cablevision. The judgement also meant that any private radio, television or cable licence could be challenged by competing applications when its licence expired.

The decision caused a sensation. The Dubé judgement, as it came to be called, was classic in its reasoning:

Neither [Victoria Cablevision Ltd. nor Capital Cable Co-operative] have a vested right in a broadcasting licence, but in my view both have a right to be heard. To be sure, the former, if he has complied in all respects with the terms of its present licence, has a priority right to be heard, but there is nothing to be found in the Act to the effect that the latter should not be heard at all. In my opinion the CRTC has a duty to hear his application before renewing the licence. Surely the additional input can do no harm and the CRTC still remains free to decide as it chooses.

One may even suggest that more competition would greatly assist the CRTC in achieving its objectives, namely "to safeguard, enrich and strengthen the cultural, political, social and economic fabric of Canada" as enunciated under Section 3 of the Act. Should the CRTC renew, without hearing other applications, it may discover too late that better and more acceptable alternatives have been passed by, perhaps to the detriment of the people in the area to be served.

It is contrary to the basic principles of natural justice to decide without hearing. True, it is just and fair to grant a licence holder priority hearing in order to decide whether his monopoly should be extended for a further term, but it is no less important that other applicants for the same licence be given the opportunity to offer alternatives: the test is bound to produce higher standards.

The judgement was released at 9 A.M. on February 3, half an hour before the CRTC was to start the second day of its three-day Vancouver hearing. The session never reopened. Boyle and the rest of the commission were so discombobulated by the judgement, on top of the previous day's proceedings, that they folded their tents and high-tailed for home.

"I have to take my legal advice — I'm not a lawyer," Boyle explained.

"The fact is," wrote Vancouver *Sun* television columnist Don Stanley, "that the CRTC has gone home to sulk." The rude suggestion of Simma Holt's lapel buttons came to pass, although not in a way she had anticipated.

The CRTC's retreat to Ottawa doubled the sensation.

The Dubé judgement brought some editorial writers to explore the issue of the right to apply. The Vancouver *Sun,* in a long lead editorial entitled "Over-ruled rules," wrote, in part:

Mr. Justice J. E. Dubé ... has thrown into question the procedures by which a large number of government commissions wield tremendous power over the lives of Canadians. ... [The judgement] is not a judgement on the relative merits of the rivals for the licence. ... [It is] a statement, in effect, that the rules [that the CRTC] has the power to set under the Broadcasting Act may, in practice, actually interfere with the fulfilment of its broader responsibility to enrich and strengthen this country. ... We would say the judge's reasoning is impeccable.

The Victoria *Times* wrote:

The Federal Court of Canada ruling in favour of Capital Cable Co-operative's argument on the renewal of Victoria Cablevision Ltd.'s licence to operate creates, as the jock broadcasters say from stadium press boxes, a whole new ball game for the Canadian broadcasting business. ... Judge Dubé's ruling is not the ultimate deciding factor in the case, of course ... But it has stirred up a healthy debate, prodded a complacent CRTC which too often just rubberstamps renewal applications, and put on guard Canadian broadcasters who have come to regard their licences as valuable private property.

Jean-Claude Leclerc, in the *Bloc-notes* section of the editorial page of Montreal's *Le Devoir,* discussed the CRTC at length, as well as the judgement. The analysis, the most perceptive of them all, was entitled "More important than the French language":

> If Quebeckers escape in the coming years the profitable silliness unleashed on the airwaves with the blessing of the CRTC, they will owe it, perhaps, to a modest group of activists in British Columbia. There was something in effect more important than French-language television to settle in Vancouver, namely the right of citizens to challenge the lucrative monopolies that the Canadian Radio-Television Commission has established across the country. The thing has been done.

Capital Cable Co-operative now having applicant status, its *modus operandi* could change. It was now worthwhile to make an expenditure of more time and money and to fully develop the licence proposal. It was also possible to approach the public with the real possibility of becoming the licensee. Immediately following the Dubé judgement, Capital Cable held its first general membership meeting, in the Empress Hotel in Victoria. It was heavily attended. An extensive committee system was established and went to work. A storefront office was opened on Yates Street, a few blocks from downtown, in a building owned by Peter Pollen Ford. The APBBC provided small honoraria for two people. Everything else was donated or paid for by Capital Cable membership contributions. Circulars spelling out every aspect of the cable battle and the Capital Cable licence proposal — rates and ownership, programming, program advisory council, local ownership, how community-owned cable systems work (the Campbell River T. V. Association), the legal story — were issued. These circulars included a comparative analysis of the existing holding-company operation. By this time the co-operative's membership was approaching the five-hundred mark. New sponsors had joined the list. It included, as well as Pollen, Young and some other business people, leading cultural personalities, academics, two school-board members, the secretary-general of the Victoria Labour Council (soon to become an alderman), a major local architect, a few journalists, writers and lawyers, and other well known and not so well known citizens. Some of them, like Hugh Keenleyside, retired journalist Jack Scott, Walter Young (chairman of the Department of Political Science, University of Victoria) and Murray Adaskin (composer and teacher) were nationally known in their fields.

One of the first things the Capital Cable office did was to fight, in May 1976, a rate-increase application by Victoria Cablevision, which was asking for a hike from $4.50 to $6.00 per month in the basic subscriber charge, despite its excessive rate of return at the old rate. In the previous financial year Victoria Cablevision had spent $83,000, or 3 per cent of subscriber revenue, on the community channel. In the rate-increase application filed after the Dubé judgement, the listed anticipated expenditure on the community channel dramatically quadrupled to a purported $363,000, or 12.4 per cent of subscriber revenue. The new figure was not all that it appeared to be. It included capital expenditures, including some that had taken place the previous fiscal year. The true comparative figure — operating cost for the community channel without depreciation — came to approximately 4.8 per cent of subscriber revenue, well below Capital Cable's projected 10 per cent, but nevertheless something of an improvement over the previous year. Similarly, in the pre-Dubé renewal application, where community programming should have been of major concern, only one nominal paragraph was allocated to it. In the post-Dubé rate-increase application, where the financial return was supposed to be the major consideration, the community channel became a major preoccupation. Twenty-three pages of the application were devoted to it. The prescription in the Dubé judgement that the test of competitive hearings "is bound to produce higher standards" was thereby proven by Premier Cablevision, without the competitive hearing yet having taken place!

Now, however, the process went into reverse. In early April 1976, just before the rate-increase hearing, the Federal Court of Appeal overturned the Dubé judgement without giving reasons (courts, it may come as a surprise to lay people, do not have to explain themselves). The Capital Cable office was closed shortly after and never reopened. The information materials, the blank shareholder certificates and the symbolic Capital Cable sign were put into storage. The Supreme Court subsequently denied leave to appeal, bringing the court action to an end.

The renewal hearing for the licence, postponed by the court case, followed in January 1977. The courts had decided only that the CRTC was not required by law to hold competitive hearings in these cases, not that the CRTC could not or should not hold such hearings. But the CRTC's machinery had already begun to downgrade the matter. Only two of the nine full-time commissioners (the members of the executive committee with voting power) were at the hearing. Chairman Harry Boyle was significant by his absence.

The Capital Cable brief to the commission spelled out again the fundamental principle of competitive hearings and all the other arguments. It outlined in detail the Capital Cable viewer-owned licensee proposal. It pointed out how its objectives coincided with the CRTC's stated objectives, while Premier, interested in spending as little money as possible on programming, faced another direction. It underlined the importance, for reaching Canadian objectives, both of competitive hearings in general and of the Capital Cable initiative. It also suggested that given the monopoly nature of cable licences in any one territory, the commission could introduce competitive hearings first for cable, leaving radio and television until later.

The hearing was the first direct encounter with the CRTC over the issue of the right to apply, and the first time the CRTC was obliged to discuss publicly the substance of the issue. Capital Cable got down to the essentials. With Victoria Cablevision's financial statement from the rate-increase application in hand, Capital Cable had commissioned the accounting firm of Thorne Riddell & Co. to draw up detailed financial projections for a ten-year period, based on Capital Cable operating specifications and the cost of purchasing the system. These projections, filed for the renewal hearing, were a revelation. They were based on the old rate of $4.50 for individual subscribers; a rate adjustment upward for inflation was made after the first five years. The projections showed that a Capital Cable licensee would mean for Victoria, from a medium-sized cable system approaching eighty thousand subscribers, a minimum annual programming expenditure by 1986 of $1,723,000. Half of this, or $861,500, would go to new professional program production. At a more than generous $5 million compensation price (twice shareholders' equity — that is, twice proper compensation in a regulated industry), another $1 million annually from surplus revenue would become available for programming. If one extended the capacity of a Capital Cable licensee to the country as a whole, and Capital Cable always had the evolution of all Canadian television firmly in mind, the implications were even more interesting: a new sector of professional television and film production, down the road, approaching the $100-million mark.

There was no doubt now that the Capital Cable licence proposal outweighed arguments for reverting to the past licensee. The margin of difference was too great, and it was clear-cut and structural. A CRTC staff member present for the hearing even felt obliged to volunteer the judgement. A representative from the Department of Communications, sent out to monitor the hearing, went away convinced there should be a competitive procedure.

Capital Cable also had outlined in its submission the process of negotiation for acquiring the assets of Victoria Cablevision, should the licence change hands. When the question was put at the hearing, Premier agreed to sell the assets of Victoria Cablevision to a new licensee should the licence be reallocated. Premier also agreed to provide a competitive hearing with evaluations of the assets based on different criteria. Capital Cable had suggested the same method, offering to put forward, in its licence application, its criterion for fair compensation. The issue would then be discussed by all interested parties at the hearing, after which the commission would be able to advance guidelines for negotiations. Or the commission, if it preferred, could leave the negotiations completely to the two parties. Capital Cable also suggested an arbitration process if necessary.

There, finally, it all was. No procedural or administrative obstacles — no practical obstacles of any kind — stood in the way of a competitive hearing.

Another factor had cropped up since the court action. The principals of Premier Cablevision were not even intending to keep the Victoria cable licence or the other licences they controlled. On the very day the renewal hearing was held, they announced an agreement to sell their controlling interest in the company to Western Broadcasting, who were corporate friends, and, in the process, to hand over the licences to them. (Previous to that, the principals of Premier had an arrangement with Rogers Telecommunications of Toronto to sell their controlling interest and to hand over the Victoria licence and the others to Rogers. This application was to have been heard at the same time as the renewal application, but was withdrawn.)

The CRTC's closed procedure when the existing licence expired was objectionable enough; but to exclude competing applications when the licence was being reallocated to an altogether different party was ludicrous. The commission, at the renewal hearing, was entertaining an exclusive application from a holding company which wanted the licence only to traffic with it to make a capital gain, while denying the right to apply to the one party in the proceedings that wanted the licence for the reason the licence existed.

In its decision the commission took the issue of a competitive procedure "under advisement." It just refused to decide. At the same time, instead of granting a short-term renewal, during which period the commissioners could finally, belatedly, sort the matter out, it blithely renewed the licence for another four years. No comparison of the two licence proposals was made, either, although everybody knew that contention between the two licence proposals had been the real issue at the hearing. Even one of the commissioners, after a heavy-handed attack by Premier on Capital Cable, was brought to note aloud that it seemed as if Capital Cable were the applicant and Premier were the intervenor.

Next, after the CRTC, it was the turn of Jeanne Sauvé, the minister of communications. The Governor-in-Council, that is, the Cabinet, is empowered by the Broadcasting Act to set aside or send back for reconsideration any decision by the CRTC. Capital Cable duly filed a lengthy appeal.

In ordinary licence decisions, the Cabinet judiciously avoids intervening, even if it is petitioned. To set aside those CRTC decisions would be duplicating the licensing function of the commission, undermining one of the legitimate reasons for having established the agency. This case, though, involved a crucial question of democratic principle and was right at the heart of the licensing process and implementation of the legislation. It was exactly the kind of case for which the powers of the Cabinet were established.

Capital Cable's petition was denied. "The decision," wrote Sauvé, "was based primarily on the fact that if your request had been granted, it might have compromised the independence of the CRTC. . . . The questions at issue were largely procedural matters which should be treated as falling within the responsibility of the commission." The letter concluded that "nonetheless, I believe that the points you have raised are sufficiently forceful to warrant more thorough examination. To this end, we will be taking a further look at the whole problem in the near future." In other words, we have thrown out the case without really knowing what we were talking about or what we were doing to the broadcasting structure, and now that we have, we are going to find out what we should have known.

This was administration of the country at its highest level. There was no provision in the statute for appeal to God.

3

The Principle of Membership-Governed Licensees

THE IDEA OF membership-governed licensees in the Capital Cable licence concept had a particular importance of its own.

As long as membership-owned and governed licensees were restricted to marginal radio stations, and northern and rural operations, and to a handful of fragmented cable territories, they could always be discounted as exceptions and as somehow exotic. They would also lack the organizational and political muscle that comes from a large number of like-minded, financially healthy organizations banding together in a trade association or federation. The one chance of such an integrated co-operative structure, in Saskatchewan cable, was destroyed by the CRTC with its 1976 decision.

Except for a few history aficionados, everybody had forgotten, if they ever knew, that the Canadian Radio League had once hoped that co-operative and community radio stations would form a substantial part of local broadcasting. There was, though, an existing model of a membership-governed sector in broadcasting, in the Netherlands, where membership associations were the base of the structure and did most of the broadcasting, and had been doing so for fifty years. They *were* the system.

The membership structure in the Netherlands went back to the early radio days, when associations representing the main religious and political elements in Dutch society, the "pillars of society" as those elements were called, pioneered radio broadcasting. These five radio organizations, Catholic, Protestant and Socialist, constituted the entire structure up to and through the early television age. In 1965 provision was made for new organizations to qualify as full-fledged broadcasting associations, should they gain enough members through a trial period. In the mid 1970s there were seven broadcasting associations in total, in television as well as in radio. A CBC- or BBC-type organization was also established in the 1960s, and did 30 and 20 per cent of the television and radio programming respectively, as well as co-ordinating operations at the broadcasting centre at Hilversum, southeast of Amsterdam.

Viewer-listeners joined the organization of their choice and received in return a subscription to its programming guide, which only the broadcasting societies, holding copyright, could publish; this provided an extra incentive to join one or the other of the societies. These organizations in turn shared, on a rough per-member basis, the combined air time, the funds available for broadcasting, and the use of the facilities and equipment at Hilversum. This membership-based structure gave to Netherlands broadcasting a representative democratic character unknown to broadcasting systems in Canada, the United States, Great Britain, France and elsewhere. The principle of it was quite separate from the structure of organizations like the CBC or the BBC and totally removed from the structures of ownership and control of American commercial broadcasting networks.

The only people in Canada who ever talked about the Netherlands' broadcasting system, or seemed to know anything about it, were the Association for Public Broadcasting in British Columbia (APBBC). The Netherlands model moved the discussion from "access" on community channels, with marginal audiences and resources, to ownership and control of a major broadcasting sector. The cable subscription mechanism — individual households linked together by a billing system — lent itself perfectly to the idea of a sector owned and governed directly by the viewers themselves. If you were not thinking of that possibility, it was natural to keep the Netherlands example out of mind, too.

The APBBC had first become interested in the Netherlands system because of its segmented governing structure. It allowed different self-ruling organizations to share the same channels and radio frequencies. A sector funded by the cable subscription mechanism, in such a sharing arrangement, would not have to be rich enough to underwrite the operation of a whole provincial or national network all by itself. Once it got on its feet and expanded, it could branch out to a separate channel of its own.

As the Capital Cable model developed, interest in the Netherlands model shifted to the membership-governing aspect, except that in the Canadian case the federation of organizations sharing the schedule and programming budget would follow regional lines, rather than the religious and political lines of the Netherlands system corresponding to the make-up of Dutch society. What most fed the APBBC's deep attachment to the idea of viewer-governed broadcasting was its decentralized, democratic character.

The sheer democratic value of a subscriber-governed sector made it worth fighting for. It would also be a counterweight to the ownership and control of most of the mass media by a relatively few private corporations, save for the CBC. But it would not duplicate public ownership through the crown, like the CBC. Concentration of ownership of daily newspapers was a fait accompli. Concentration in broadcasting and cable was escalating. Here, with a subscriber-governed cable sector, was a God-given opportunity for deconcentration and diversification.

Separate media ownership does not guarantee the widest range of expression when the separate ownership is embedded in the same social, financial, geographical or political setting, or where it shares the same kind of internal structure — private corporate ownership of daily newspapers, for example, regardless of concentration, or Toronto dominance of English-language television networks, regardless of whether it is the publicly owned CBC or the privately owned CTV. The maximum independence for divergent expression comes from differing governing structures which reflect different social, geographical and political arrangements. The more differing governing structures there are, moreover, the freer all media become.

One of the greatest limitations of existing Canadian mass media is that they do not have democratic representative ownership and control except indirectly in the case of the CBC, through its accountability to Parliament. A viewer-governed sector would have that representative governing structure. It would have a democratizing impact like nothing else in the country or continent.

It would be folly, too, to waste the financial capability of the cable subscription mechanism on nonmembership organizations, whether the CBC or commercial broadcasters. It would needlessly destroy the possibility of a membership-based sector — of a differently governed element. If the CBC needed more money, it should get it from an increase in its parliamentary appropriation, which already determined how it was governed.

There was another, equally vital function for the proposed viewer-owned sector. It would provide a social vehicle where interested Canadians could get together as members of society to discuss and make decisions about television in general.

Canadians were involved in television only as isolated viewers choosing from available programs. This relationship was a highly restricted and incomplete one. It did not involve any consideration of the social, cultural, financial and political effects of television, or whether other kinds of programming should be produced (like more children's programming or more Canadian programming), or what difficulties stood in the way of these kinds of production, or how programming was concocted and for what reasons. These considerations required that people get together in meetings and through organizations — organizations with independent power to chart programming directions — to learn the background, discuss the issues, have their ideas tested, and take action as a community. They were no less important than the decision about what channel to flick on after dinner — they were, in fact, more important — but there was no practical vehicle for it. Occasional newspaper and magazine articles on programming issues might surface, but they led nowhere in practice.

No wonder then that one of the social considerations of Canadian television, the presence of Canadian programming, was gradually flattened by the American industry steamroller. This condition also made it extraordinarily easy for commercial broadcasting in Canada, and particularly for the cable industry, to defend their Americanizing function. There were always viewers who were never obliged to think of television in a social way — "social" in the largest sense of the term, including the cultural and political life of a country — and who were ready to cry bloody murder when Canadian content objectives got in the way of the cheapest, broadest and most convenient access to U.S. programming.

The politicking imbalance was the same. The mythology surrounding the old Canadian Radio League — the story of a group of citizens lobbying for publicly owned radio, and succeeding — hid the reality. The league had worked in uniquely favourable circumstances with relatively few people and only for a few years, and with the help of private resources. Once the national broadcasting system was legislated, however, the voluntary league dissolved. The private commercial broadcasting lobby then went to work to boost the transmitting power and multiply the number of private stations, gradually overwhelming the publicly owned element which was supposed to predominate. This pattern — of public-interest fatigue and of well financed special-interest lobbies getting the best in the long run — is familiar everywhere.

The only public-interest vehicle that could contend in the real world of television politics would have to be an operating organization with its own large and assured cash flow, and preferably one that could make programming decisions by itself. A subscriber-governed sector was tailor-made for it. There is nothing like an operating organization, with the power and resources to do things on its own in a big way, to keep citizens interested and involved and to make their participation worthwhile. The membership-based sector, because of its decentralized governing structure, would also locate citizens' involvement as close as possible to the local level, where people lived and worked.

Without this social intermediary, involving interested Canadians at the local level in a sector of their own, the chances of Canadian broadcasting objectives making real headway were minimal. The politics and economics were loaded against it. Paper policies would be drafted in Ottawa indulging the industry lobbies and would be further chewed up as they went through the mill. Royal commissions, advisory committees, policy review committees, consultative committees and other Ottawa rituals would come and go, disappearing into the great paper shredder of history. Whatever initiatives survived would be wide of the mark and marginal. Ineffectiveness was as predictable as the sun coming up in the morning.

4

Licence Trafficking

IN THE CAPITAL CABLE licensing battle, the fact that the controlling Premier shareholders wanted to get rid of their cable licences and were trafficking in them to their friends created a terrible problem in public administration for Ottawa: not over what to do about it, but over how to explain it away. But it could not be explained away. For if the exclusion of competing applications when a monopoly cable licence expired offended people's sense of democracy and justice, the same exclusion when the licence changed hands — when a new licence, in effect, was issued — was even more obviously ludicrous. You could not buy the right to be the sole applicant for the licence when it was first issued, but by being a collaborator in the private trafficking of the licence, you could buy the right from the party that had originally got the licence. What was spent on the inflated purchase price in turn was lost to the broadcasting system (programming, lower cable rates), where it belonged. Going along with this procedure required from Ottawa absurdities and evasions that would have embarrassed a trained clown.

The futile skirmish over the reallocation (so-called transfer) affair would drag on until 1983.

The exclusion of competing applications when a licence was allocated to a new party — the allocation of a new licence, in effect — was first brought up by the Association for Public Broadcasting in British Columbia (APBBC) in 1976, after a Calgary holding company had filed an application to take over the cable licences for North and West Vancouver, Nanaimo and the West Kootenay held by an Ontario-based holding company. British Columbia organizations and companies were not allowed to apply for these licences in their own province and their own communities. Nobody else was allowed to apply, either. Not only were no competitive or alternative applications called for by the commission, but the word *licence* was not even mentioned in the CRTC's notice of hearing, although the licence was in fact the matter at hand.

The takeover application was approved. The commission, in its decision, declined to address itself to the crucial, prior issue. It simply did not mention it, even to dismiss it. The first all-purpose general technique of regulatory evasion is to avoid an issue altogether, no matter how prior and central it may be.

The exclusion of competing applications was next brought forward by Capital Cable Co-operative. In the fall of 1976 Premier arranged to sell out to Rogers Telecommunications of Toronto, which would give the Toronto holding company the Vancouver, Victoria and Coquitlam cable licences along with others in Ontario, again without competing applications allowed. By accident, through some of the work produced by telecommunications consultant Robert Babe, Capital Cable had run across a 1969 CRTC policy statement, "On the Pricing of Broadcasting Undertakings," which said, in part:

> It is ... the policy of the Commission to scrutinize applications for transfer of assets of licensees or for transfer of control of licensees in a manner comparable to its examination of applications for licences for new undertakings.
>
> Consistent with previous practice, such applications are subject to public hearings, at which objections may be raised *and at which companies or persons other than the purchaser proposed by the current licensee may apply for the licence.* [My italics.]

Capital Cable could, according to this, apply for the Victoria licence. It duly sent off a telegram to Harry Boyle asking that the takeover hearing be deferred so it could file an application.

The existence of this 1969 policy announcement was a bit of a mystery. The CRTC appeared to have put it on a shelf, hoping that nobody would come across it. The *Broadcasting and Cable Television Regulatory Handbook*, published by the Law Society of Upper Canada, had omitted it, although the author of the supposedly authoritative compendium was Peter Grant, a former special counsel to the CRTC and a well-known communications lawyer. A check with the CRTC's general counsel and its director of licensing confirmed, however, that the policy was still in force, although it had never been implemented. The CRTC's director of licensing then suggested that it was too late for a Capital Cable application because notice of the hearing had already been issued. This was a most peculiar notion because, until notice of hearing had been issued, there was no official way for other parties to know that there was an occasion for applying for the licence. The director of licensing then explained that in searching around for buyers, the seller will contact several parties, which is how other companies or persons learn what is about to happen. By inference, if some party was not contacted, it was out of luck. "That is tantamount to an admission that trading in licences within a closed circle — all in the family — is approved of by the commission," commented Capital Cable in a news release. "One wonders whether the CRTC is running a regulatory agency or a secret society."

Capital Cable took it for granted the CRTC would stonewall. The only question was which excuse it would use. The commission, in a telegrammed reply, simply declared its own policy null and void and wrote it out of the books:

> Irrespective of [the policy announcement], the commission has consist-
> ently followed the practice in all such applications of hearing only those
> companies and persons who have a valid agreement with the licensee or
> shareholders of the licensee. This practice has therefore rendered void
> that part of the July 10, 1969 announcement referred to above.
>
> It should also be noted that the Premier/Rogers application seeks
> approval for a transfer of shares in Premier. Premier is not a licensee
> but controls companies which are licensees. Where such applications
> receive approval by the CRTC no new licences are issued. The current
> licences continue as there is no change in the licensee companies other
> than ultimate control over their shares.

This response astounded even Capital Cable which, in its cynicism about the CRTC, thought it was prepared for any possible silliness. The practical reality of what was happening, as everybody knew, was that Rogers Telecommunications of Toronto could apply to take over the licence, whereas a community-based Victoria organization could not. The licence was changing hands. This is why the hearing was called in the first place.

The takeover process lurched forward. Capital Cable filed a bluntly worded intervention. Rogers, either because his financing broke down or because he did not want to reveal certain internal corporate workings, at the last moment folded the deal and withdrew the application.

The abandoned Rogers application was only the first in a wave of backdoor licence-trafficking deals. In January 1977 Maclean-Hunter Cable TV attempted to take over Western Cablevision and, with it, the cable licences for New Westminster, Surrey (the fast-growing southeast section of Greater Vancouver), and Abbotsford and Clearbrook, stretching out into the Fraser Valley. The main public defence against the privileged takeover was an intervention by the Canadian Broadcasting League, represented by the Public Interest Advocacy Centre and its general counsel, Andrew Roman. The APBBC also intervened. Allied in the action with the Canadian Broadcasting League was the Lower Fraser Valley Committee for Community Based Cablevision Services, a local spin-off of the association established for the occasion.

There are two kinds of arguments at hearings of this kind. One is the "policy," or "public-interest," argument, based on the idea that something should be done because it is right, or practical, or the best alternative; most interventions are of this sort. The other kind of argument is legal, having to do with what the agency must do, or cannot do, by law. Usually these arguments are presented in the form of preliminary motions. As well as the policy arguments printed in the interventions, Roman argued, in a preliminary motion, that the commission was required by law to hear competing applications, because what was involved was the allocation of a new licence, not a "transfer." The Broadcasting Act did not grant the commission the power to approve a "transfer" of licences, only to issue, amend, renew, revoke or suspend them. Also, licences were not normally for sale — a licence to practise medicine or law, for example, or any other licence "granted on the basis of the experience and skill of the person(s) being licensed." The "privilege granted by the licence" was personal to the licensees and not a commodity.

In questioning by Charles Dalfen, a commission vice-chairman and also a lawyer, it surfaced that in the CRTC's official view all that it was doing in these "transfer" cases was amending a condition of the licence. There was no new licence issued. The holder of the licence was simply a condition to be attached to the licence.

The CRTC subsequently denied Roman's motion. For years after, legal counsel of the commission, and the commission with them, made an exhibition of themselves by describing the reallocation of licences as just changes of condition of licences, and they did so with solemn, dogged earnestness. The lawyers for the traffickers in licences were only too glad to indulge them.

Eventually Maclean-Hunter's takeover application was turned down by the CRTC, but for other reasons. Except for the legal reference rejecting Roman's motion, which was separate and apart, the commission, in its decision, again declined to address itself in any way to the issue of competitive hearings when a new licence for an existing territory was issued. The legal issue itself went to court.

With the Rogers Telecommunications takeover application of Premier withdrawn in 1977, Premier quickly worked out another deal, this time with Western Broadcasting (not to be confused with Western Cablevision). Western Broadcasting was a large Vancouver-based conglomerate with radio, television and other holdings, including a 50 per cent holding in BCTV and radio stations in Vancouver, Calgary, Winnipeg and Hamilton. It also had a controlling interest in Northwest Sports Enterprises, owner of the Vancouver Canucks, in which Premier had a small interest. The founder and chairman of Premier, Syd Welsh, and the president of Western Broadcasting, Frank Griffiths, were long-time friends and sat together on the board of Northwest Sports Enterprises.

The CRTC again excluded Capital Cable Co-operative from applying for the Victoria licence. It was now more than two years and several other incidents down the line since the APBBC had first raised the issue of competitive hearings in these cases, to be met by commission stonewalling.

If the Victoria licence now were given to Western Broadcasting, it and the others held by Premier would be gone for another generation, unless competitive hearings were implemented at renewal time. The CRTC had already barred the door to that. To force the issue, Capital Cable made a formal cash offer to Premier for the Victoria system, competitive with Western Broadcasting. It did so under protest, vigorously insisting that the licence should be open to applications from all interested parties and that private trafficking in public licences was abhorrent.

The offer was for $4.5 million. It was based on the ratio of Victoria cable subscribers to Premier's overall subscriber total, then applied to the Western Broadcasting offer for Premier as a whole. The Capital Cable offer was almost 25 per cent, or $865,000, over the market value of the Victoria system, based on the current price for Premier shares divided by the per-subscriber ratio. It was double the shareholders' equity in Victoria Cablevision, the proper measure of compensation in a licensed monopoly situation. Premier summarily turned down the Capital Cable offer.

Capital Cable, as part of its intervention, filed the Thorne Riddell financial projections showing how it could easily undertake its major program production initiatives and also pay for the acquisition of the system and run up a large cash surplus, all at the existing subscriber rate. Capital Cable also gave the commission notice of legal action if it approved the privileged Western Broadcasting application. The organization wrote angrily, in its submission:

> The real issue here is the irresponsibility of your commission itself.
>
> You are willingly hearing this undesirable application for the merger of two communications giants — giants who have offered not one single advantage to the public, to society at large, as a consequence of their proposal — and yet you will not even hear an application from Capital Cable in Victoria for one of the very same licences involved, the cable licence in their own city.
>
> This practice is . . . reprehensible. . . . Licences are granted openly in competitive circumstances, and a new licence for the service should not be issued to private friends of the outgoing licensee — for that is what is involved here — without the hearing of competitive applications from other interested parties.
>
> By allowing the outgoing licensee to limit the field of potential new licensees — to choose within limits its own successor — the commission surrenders an important part of its responsibility.
>
> By insisting upon the exclusion of other applicants for what is in effect a new licence, the commission practices barefaced favouritism, and illustrates again, in the process, its dominance by the industry it has been established to regulate in the interest of the public. To borrow a phrase (from Blaik Kirby, *Globe and Mail* . . .), "the CRTC does not serve the public good nearly so energetically as it serves the broadcasting establishment."

The Capital Cable attack was stepped up at the hearing itself. Chairman Harry Boyle did not have a pleasant time. He was unhappy to begin with in this Vancouver place where he had run into so much trouble already.

But did this strategy do any particular good? Of course it didn't. Although the commission rejected the Western Broadcasting takeover, it once again, in its decision, simply refused to discuss in any way the issue of competitive hearings. On the other hand, nothing else did any good, either, so one might as well not pull any punches. Letting the commission have it between the eyes was honest to events and left the press and the newspaper-reading public without illusions about what the CRTC was doing.

But let's trace the CRTC's captivity back to its tether stake. The hearing on the Western Broadcasting takeover attempt and other CRTC hearings on takeover applications only nominally had to do with choosing a new licensee. The licence trafficking was their real function, and the CRTC's captive role was to ensure that the trafficking was allowed to go on. The CRTC was The Connection.

There are two elements in the trafficking price: just compensation for the investment and the trafficking margin. In a competitive hearing the outgoing licensee would receive the just compensation regardless of who was awarded the licence, making the selection process financially irrelevant to the outgoing licensee. It had no business in the first place limiting possible applicants and interfering in the licensing process. The CRTC, by the same token, would freely choose from all contenders able to meet the compensation price. But, exactly in so doing, the commission would eliminate the trafficking margin.

To allow the trafficking payoff, the CRTC had to exclude competing applications.

"The maverick czar of Canadian communications does not suffer bureaucrats gladly," said the subtitle of a wonderful article called "What Makes Harry Boyle?", by Silver Donald Cameron, which appeared, during this period, in *Weekend Magazine*. Boyle was "blunt and unpretentious, shrewd and sentimental . . . maybe the only unembalmed Canadian in town," the article gushed. Cameron and Boyle together dissected Ottawa — "a government so smug and remote that it scarcely seems to know which country it's running," in Cameron's words; "Yeah," nodded Boyle — and established just how folksy and un-Ottawa the CRTC chairman was. It was an inspiring and touching portrait. Cameron did not get as far as Saskatchewan to record Boyle's and the commission's derailing of the cable co-operative movement and the misconceived attempt to make an end run around the Saskatchewan government. Nor did he make it to British Columbia. So the events in the battle for competitive hearings also did not get into Cameron's copy.

The maverick of yesteryear was not so maverick any more. Boyle had become a reverse image of the original. "Compromise like bureaucracy spawns self-preservation," he had written to a friend in 1965. This was as good an epigram as any to catch the later Boyle himself, except that "evasion" needed to be added to "compromise" in the simile.

Boyle's explanation for not allowing competing applications — "It's just not done that way" — and the gobbledygook attempt to explain away the change of a licensee as not being a change of licensee were just the kind of stonewalling game that made other people not suffer bureaucrats gladly. The commission's uptight manoeuvres in Saskatchewan were a caricature of the uptight Ottawa that Boyle was supposed to have so much disdain for.

In a speech in 1969, when he was CRTC vice-chairman, Boyle declared that "to use the hardware [of broadcasting] for propaganda purposes will merely hasten the millennium." Yet when powerful corporations began to do just that with "advocacy advertising," in defiance of the CRTC's own policy, and the matter was brought to the CRTC's attention, Boyle, first as vice-chairman and then as chairman, declined to stand up against it.

A true enlightened maverick, in an administrative agency like the CRTC, would be somebody who adheres to general principles with toughness, openness and clarity. An enlightened and also effective maverick in that situation is somebody who not only adheres to general principles but also keeps track of them in a sea of trivia and complications. Avuncular Boyle, alas, was never that maverick. He was no match for the bureaucratic law that trivia takes precedence over important matters. He was, a senior staff member of the period recalls, "a fine, thoughtful man, he had enormous heart, he had a feeling for broadcasting and culture, but in the job, he lacked logical toughness, he would often take the easy way out which tended to favour one group."

When Boyle joined the fledgling commission in 1968, and during its first few years, he was still the fighter, the unconventional anti-bureaucrat, the romantic idealist — the same man recalled with admiration and affection by the figures from the past that journalist Cameron talked to. He had few illusions about the commercial broadcasters. The behind-the-scenes mudslinging campaign about Juneau and himself by the commercial broadcasting lobby, during the Canadian-content drama in 1970, so astonished him in its nastiness and slander that it was all he could do to contain himself from letting them have it head-on. Their crass commercialism and shallowness — above all, their limited feeling as Canadians — he found beyond imagination.

By the summer of 1972, after a lot of soul-wrestling, he had privately decided to wind up his days at the commission. He seemed to have realized that his hopes for the commission were a snare and a delusion. But instead of leaving the comedy after all, he hung on and became part of it. In January 1976 he accepted the chairmanship, even though a secret arrangement was attached denying him the statutory security of the full four-year tenure officially announced. "I just put my own feelings aside," he explained, "and said it's just one of those things. . . . It didn't please me." No explanation for the secret condition was ever forthcoming. He resigned in 1977, less than two years into his appointment, because he was given no choice.

5

How Working
Papers Work

IT IS OFTEN THROUGH complicated documents and dull, out-of-the-way, little-noticed hearings that an agency can ill serve the public interest in a big way.

On July 25, 1978, the CRTC issued a working paper entitled *Proposed CRTC Procedures and Practices Relating to Broadcasting Matters.* It was forty-eight pages long; its table of contents listed twenty-three subjects. The commission invited public comment and scheduled a hearing in Ottawa for November, after which it would "make its findings and . . . again [invite] public comment." Among the major stated purposes of the exercise were "to facilitate a wider and more informed participation by the public in the broadcasting regulatory process" and "to ensure the fairness of commission proceedings for all parties."

There was no provision for costs to help interested citizens prepare submissions or travel to Ottawa for the hearing. The subject of costs to intervenors for hearings in general also was not mentioned in the impressive-looking working paper. The hearing was held during office hours, as usual, in a small, conventional meeting room on the twentieth floor of an office tower on Laurier Avenue. Of the twenty-seven presentations at the hearing, twenty-one were from the private broadcasting and cable lobbies or their members. Of the remaining six, three were from public-interest organizations, one from an academic specialist in communications, one from a performers' union, and only one from an individual member of the public. He came from Hull, just across the river. It was a hearing on public participation without the public participating.

The hearing was a traditional meeting of the CRTC and private broadcasting and cable representatives. The crisp, well-dressed uniformity in the meeting room gave it a clubby businessmen's atmosphere. The broadcasting lawyers were also out in force, including two who had been with the commission, one as a commissioner (John Hylton), one as general counsel (Chris Johnston). There was comforting familiarity and shaking of hands all around, although the tedious proceedings in the claustrophobic, smoketinged room were not inspiring.

Because most of the important members of the club had turned out to make an appearance, the commission was faced with a long list of presentations, which it appeared to be trying to get through as quickly as possible. CTV president Murray Chercover caught the spirit of the gathering exquisitely when he told the commission that the public did not need to contribute through a larger involvement in the regulatory process. The CRTC would look after the public interest. Broadcasting issues had become too complex for the public to contribute more effectively anyway. Besides, more public involvement would "create a new gulf between the regulatory agency and the licensees" and interfere with the "intimate" relationship between the two. There was in all this procedural talk — a recommendation for disclosure of financial information, for example — the disturbing suggestion of an adversary relationship, Chercover pointed out, whereas the licensees and the CRTC had "a joint and mutual responsibility" that was "a consultative function."

The commission's legal counsel, Jack Johnson, who did have a reformist bent of mind, gave in to temptation. "I guess that almost begs the question," said Johnson. "Why a hearing?" The hearing was to allow a handful of subjectively interested groups like ACTRA to appear, replied Chercover, but otherwise he could not think of any reason.

The real importance of *Proposed CRTC Procedures and Practices Relating to Broadcasting Matters* was that for the first time the CRTC attempted to explain why it would not allow competing applications when licences expired (renewal time) and when licences for existing territories were allocated to new parties (so-called transfers). It was two and a half years since Capital Cable Co-operative had raised the first question, and approximately the same length of time since the Association for Public Broadcasting in British Columbia (APBBC) had raised the second question. The commission, through all this period, had postponed no decisions that might be affected by these issues and blithely proceeded as it always had.

The *Proposed CRTC Procedures and Practices Relating to Broadcasting Matters* was a ranking contender for one of the shoddiest documents in the history of the Western world. The document, on analysis, gave every appearance of the commission's having determined ahead of time that no competitive hearings would be allowed and that reform should be sidetracked; then it gathered up any conceivable and some inconceivable arguments, variable, contradictory and irrelevant as they might be, to pass off the predetermined conclusion. That was more or less what happened.

The principle of competitive hearings for the use of public property wasn't discussed. There was no searching democratic voice, or even faintest whisper, in the document when it came to competitive hearings. The document also failed to make the basic and crucial distinction between the value of the public licence, which should always be retained by the public, on the one hand, and private value reflecting net investment in assets, which is properly assignable to the licensee, on the other hand. The paper omitted even to bring forward for discussion the notorious trafficking in licences.

The fundamental practical advantages of competitive hearings when existing licences expired — the best possible licensees and a disciplining effect on existing licensees — were not explored. Similarly, practical illustrations of competitive hearings or of models for competitive hearings were not put forward. The prominent existing example — the competitive reallocation of television "franchises" in Great Britain — was not considered. The Capital Cable licence challenge — in effect a dry-run competitive hearing — was not examined.

Instead of establishing models for competitive hearings to explore how they would work, drawing on real, practical situations, and instead of a clear and measured reasoning process as to the pros and cons of competitive hearings, what the document offered was a vague collection of random considerations and concepts without explanatory context, followed by a "having regard to all of the foregoing." None of these considerations was explained or justified in any practical context. Some were even listed as questions, with no analysis of whether the questions were legitimate or not (most of them weren't) and no inquiry into them.

For example, the legal history on the issue was cited as a consideration. But the court case established only that the commission was not obliged by law to consider competitive applications. It did not say that the commission was unable to do so or that it should not do so. The reference was quite specious. (If the court had decided otherwise, the issue would not be up for discussion!)

The document cited the "fact" that long-term investment and loans would not be forthcoming and that there would be an unhealthy pressure on licensees to maximize short-term profits if they were unable to expect that good performance would normally result in renewal. But this was not a fact at all. It was also a feeble probability. In the competitive procedure as proposed, fair compensation was guaranteed. Nobody would be left in the lurch. Nothing would get in the way of long-term investment and loans if used for legitimate purposes.

As things were, licensees in Canada maximized short-term and long-term profits where it suited them, and maximized the speculative gain in the private trading of public licences, all to the cost of the quality of service and to subscribers. They did so with relative impunity exactly because no competitive licensing procedure was allowed. (A quick glance at Great Britain and its competitive television licence procedure would have shown that television "contractors" both allocated sufficient investment and were under strong inducement to meet the regulatory authority's programming objectives.)

The commission's document referred to an alleged "vexing question of how, in a competitive licence renewal hearing, an incumbent licensee with a 'track record' and a new applicant without are dealt with on an even-handed basis and subject to the same criteria." This was a question of sorts, but for the new applicant rather than for the commission. He would have to convince the commission not only of the benefits he would bring to the licence but also that they were attainable.

The authors of the document wondered "whether simply implementing a competitive renewal process might alter only the form and not the substance of the process and not really grapple with the need to make the renewal process more effective and meaningful." But the one, of course, did not preclude the other. The document raised the alleged "question" as to "whether if a competitive renewal process were in effect and if, in particular cases, there arose no competing applications, there would be almost irresistible pressure for automatic licence renewal; in other words, whether such a system can be effective in the absence of competing applications." Again, there was no connection between the one and the other.

These "questions" raised by the CRTC simply did not apply. What did apply was the cynical inference, here in the commission's own document, that the commission would undermine a competitive procedure by not applying itself vigorously to make it work.

The commission's document made several references to the wide latitude given intervenors, whereby they can "point out how a superior performance of greater benefit to the public interest could be obtained if a different party had the licence. In brief, existing rules do not inhibit comparisons." But when the document got around to listing what the major considerations would be when a licence expired and came up for reallocation, comparison of alternative licence proposals was not one of them! And if a competitive comparison were to be a major consideration, why not do things properly and allow new licence proposals the status of applicants? Without that, any comparative procedure would be highly prejudiced. Nobody would bother with a competing proposal of any substance in a totally rigged game like that.

Having covered over the practical benefits of competition (by not talking about them), avoided the matter of trafficking in licences, avoided practical cases or any other check on abstraction and artificiality, and tossed in inapplicable references, nonfacts and gratuitous speculation, the document then proposed not to allow competing applications. Even then, how the document arrived where it did from what went before was a mystery. The conclusion had been tacked on to the assortment of dilatory reflections like a skin graft attached to a body with Scotch tape.

The next part of the document dealt with the allocation of new licences when the incumbents were getting out of the business (so-called transfers), and it was still more astonishing. The report allowed that there was something wrong with the existing procedure. Licences being transferred to new parties were originally allocated "after a public hearing at which all interested parties are heard in competition . . ., and after consideration by the commission not only of the promises of performance proposed by each applicant but also of the quality, character and ability of the controlling personnel behind each applicant." Now there was only one applicant chosen by the "vendor." Although the commission retained the power to approve or deny, "denial is often an invidious choice in that it may mean leaving control of a licensee in hands which may no longer wish to operate it, and not knowing whether there are any other prospective purchasers and whether any such purchasers would be more or less suitable in terms of the public interest." So the commission, the document explained, gave serious consideration to implementing a competitive transfer system. "The commission finds, however, that while there is much merit in theory in such a process, there are also such formidable obstacles to its implementation as to render it impracticable." A change was not "feasible."

But what were these "formidable obstacles"? The CRTC argued that if the controlling block of a multiple-shareholder licensee company wanted to get out, then obliging the company to surrender the licence for purposes of a competitive hearing might be most unjust to the noncontrolling shareholders. In fact the licence was, for practical purposes, awarded to the controlling block in the first place. It was their "quality, character and ability" that was the governing consideration in the allocation of the licence, as the document itself so well described. The others were just along for the ride. This is why the commission was holding the "transfer" hearing in the first place. As several astonished observers pointed out, either it was a change of the real licensee, in which case other applications should be heard, or it was not a change, in which case the CRTC should not be bothering at all.

The noncontrolling shareholders in a licensee company did have a moral right, when the licence was surrendered, to a fair share of the compensation for assets from the new licensee and that, of course, would be forthcoming on exactly the same basis as the compensation to the controlling block. They also had the right to get together, buy up a majority of shares (or form a new controlling group where they already had a diffuse majority), name a new board and management, and file an application in competition with all other interested parties. Their interests were in no way abridged. So there was no practical or moral problem there at all, much less a "formidable obstacle."

That wasn't all, though. The document then stated that, in some cases, the control block of a licensee holding company "may well not be large enough to carry the often necessary vote of shareholders to approve the surrender of the licences." What could the commission possibly do then?

This red herring was a deep scarlet, odiferous one. Any exit by the controlling block or any definitive change in the controlling block (the effective licensee) would be a de facto surrender of the licence as previously allocated, in which situation a competitive hearing for a new licence would be called for. If there were any legal doubt about this, all the commission had to do was make control by the controlling block a binding condition of licence. Once this block surrendered control, or its composition was altered past a certain point, the licence would become null and void, automatically bringing on a hearing for a new licence.

The "formidable obstacles" did not exist. As communications economist Robert Babe would write later, commenting on the document, "simply because shares in public corporations trade in the marketplace and control over corporations can be transferred in this manner does not dictate that broadcasting licences be so traded unless the CRTC wills this to be the case." No licensing agency in any case should allow itself to become the prisoner of the corporate form of some of its licensees.

At least, however, the document and the scheduled hearing to discuss it put the matter on the table. Despite profound cynicism, the board of Capital Cable decided, after considerable discussion, that it would participate in the review. It drafted and revised a comprehensive brief on the licence "renewal" and "transfer" procedures, critically dissecting the commission's document in broad terms and then going into a section-by-section, sometimes paragraph-by-paragraph, analysis. A delegation was duly sent from Victoria to Ottawa for the hearing.

The Public Interest Advocacy Centre (PIAC), represented by general counsel Andrew Roman, also submitted an extensive brief, detailing, among other things, the administrative mechanics of competitive hearings. Liora Salter of the Department of Communications at Simon Fraser University, and one of the founders of Vancouver Co-operative Radio, attached to her brief a lengthy paper exploring competitive procedures and their advantages. The Consumers' Association of Canada called for competitive hearings.

At the hearing, however, it seemed that members of the commission had not read all the submissions, or had not read them all carefully. A couple of the commissioners, including vice-chairman Charles Dalfen, the commission's version of a bright thinker, repeated the old bogey about the risk to investment.

There was more of the same intellectual shoddiness, and glimpses of predetermined minds, on this subject of competitive hearings. The CRTC's clientele and their associates followed one after the other, except for the occasional outsider like Capital Cable, the PIAC and Professor Salter. The hearing rambled along its desultory way to an end, and the coat-check lady handed back the last of the coats.

That was it. The commission did not "make its findings and publish drafts of all proposed rules or amendments to rules, and again [invite] public comment," as the working document had promised. Nor did it respond to the points made in the submissions by Capital Cable and the PIAC, and has not responded to this day. The commission did nothing, except to continue excluding competing applications, with the attendant consequences. The working paper, as a senior staff member later explained, disappeared "into the bowels of the commission."

Well, that wasn't quite all. Years later, when challenged again about the exclusion of competing applications, the commission would refer to *Proposed CRTC Procedures and Practices Relating to Broadcasting Matters* as if it meant something and as if simply naming it was sufficient answer to any challenge. And that really was it.

The leading commission actor in the 1978 hearing on procedure was vice-chairman Charles Dalfen. For aficionados of the CRTC human comedy, or of the human comedy *tout court*, Dalfen's performance did not go unnoticed.

Dalfen was the rising star, or had been the rising star. Born and educated in Montreal, with degrees from McGill, Oxford and the University of Ottawa, Dalfen was director of the legal and regulatory branch of the Department of Communications while still in his twenties. At the age of twenty-nine he became a full-time professor of law at the University of Toronto. Two years later he was appointed deputy minister of transport and communications in the British Columbia government. During his B.C. sojourn, he made a submission to the CRTC on behalf of the British Columbia government and so impressed Harry Boyle that when the New Democratic Party lost the provincial election in late 1975 and Dalfen was looking for something new, Boyle reportedly recommended him and fought for his appointment to the commission. The appointment was made in 1976, when Dalfen was thirty-three. Boyle in his wisdom regarded Dalfen as somebody who could speak for British Columbia within the commission, although Dalfen's stay in the province was short, bureaucratic and rootless.

Dalfen, in the Camu years, was seen as the knowledgeable heavyweight at the commission — modern, bright, capable of mastering a welter of details, fluent, probing, analytical. As vice-chairman he showed a strong interest in procedure, one of the few members of the commission — maybe the only member of the commission — to do so. He had quickly moved to make public participation easier and more effective when it came to telephone regulation, which had recently been taken over by the CRTC from the Canadian Transport Commission. His rapid rise and sense of his own intelligence gave him a confidence in his opinions. His verbal ability, legal training and mental facility gave him the skills to defend them. Altogether he had the appearance of somebody who knew what he was talking about, somebody with special authority. The fact that he had gone so far so fast contributed to the impression.

So there, at the 1978 hearing, was Dalfen, faced with the proposition of real reform. His response, as he struggled with the subject, was like a man, wanting to be well dressed, taking off his clothes in public as he attempted it. The more Dalfen opened his mouth, the more he showed how captive he was. He repeated some of the shoddy, inapplicable arguments of the working paper (arguments for which he may originally have been responsible) and the assumptions contradicted by reality, as if nothing had been written about them in submissions. He wondered, in the circumstances, if the notion of competitive hearings had any soundness at all. He dragged in the court decision in support of his own thoughts, although, as had been pointed out, the court decision was on a legal point not in question at the hearing. He drew a questionable inference from the Capital Cable licence challenge, after the organization had appeared and could not comment (when the organization *had* appeared, he had declined to ask it any questions at all).

When Professor Salter made the innocent and unexceptional suggestion that competition at renewal time, and a little insecurity for existing licensees, might add needed discipline to the system, Dalfen responded vigorously: "We need nurturing and security, Miss Salter, not insecurity." "Well said," chimed in the other vice-chairman, Jean Fortier, with matching indignation.

The matter of the allocation of a licence to a new party — a licence that had originally been awarded in a competitive hearing — was even more of a dilemma for Dalfen. He had to admit there was something wrong with it. He avoided, however, addressing the key issue of the trafficking. Facing up to the issue would have put the commission in an embarrassing position, since the only function of excluding competing applications was to allow the trafficking to go on. He talked instead of how "certainly informally broadcasters have said that they are coming to the realization that they ought to present to the commission the fact that they did canvass alternatives ... the commission has expressed concerns and I think there has been sensitivity in the industry to that concern." Andrew Roman, who was appearing at the time, could hardly believe his ears, since the only alternative the trafficker was going to choose was the one that was going to pay the highest trafficking price, and he would take his chances on it. But there was something more basic and obvious. Canvassing alternatives was the responsibility of the licensing agency, not the outgoing licensee, which had no business in the matter (fair compensation for investment being assumed). And, if there was going to be such canvassing, should not all aspiring alternatives have the right to come forward to the licensing agency, and should they not all have equal procedural rights and the chance of being allocated the licence? Otherwise how could they be dealt with fairly and how could the best new licensee be chosen? Similarly, trafficking in the licence on any scale was wrong. The commissioner wasn't making sense.

It was a bit like Joseph Conrad's *Lord Jim*. A singular test of judgement arose but the bright young officer, a measure above his colleagues, the one person who should have know better, went along with the others and slipped away, leaving the vessel and its charges adrift.

But that is as far as the analogy goes. Instead of expiating any guilt by good works and death in darkest Patusan, Dalfen established a law practice in fairest Ottawa-Hull in association with two communications lawyers.

6

The Unfriendly Takeover

IN THE MIDST of the working-paper ritual, Rogers Telecommunications, the holding company of Ted Rogers, the owner of a cable system in Toronto, stole away the licences of a multisystem holding company, Canadian Cablesystems Ltd. (CCL), by stock dealings against the wishes of CCL's management, the effective licensee. This was done with CRTC approval. It was not even a "transfer" — it was called an "unfriendly takeover" and a "coerced takeover" by the Rogers organization itself. There had never been anything like it under the public licensing sun.

The scenario featured (a) a licensing agency that wasn't sure who controlled some of its licensees and that had to hold a special hearing to find out; (b) a company that, through outside stock dealing, bought the right to exclusive consideration for cable licences, excluding even the existing "licensee" group under licence by the commission, and (c) an applicant filing applications to which it was bitterly opposed.

The players included CCL, which held licences for eleven different systems serving approximately twenty cities, towns and areas with a total of 465,000 subscribers, including 120,000 in Greater Toronto. The other licences were for Mississauga, Burlington, Hamilton, Brantford, Paris, Newmarket, Cornwall, Kitchener, Stratford, London, Oshawa, Whitby, Bowmanville, Chatham, Kingston and Calgary (this latter recently acquired from another company after a typical noncompetitive hearing). CCL also had theatre (Famous Players) and real estate holdings. It ranked a close second in cable holdings to Premier Cablevision, which had 500,000 subscribers.

The costar was Rogers Telecommunications, owner of Rogers Cable, which had more than 200,000 subscribers in Greater

Toronto plus Brampton, as well as Rogers Radio (which had CHFI-FM and CFTR in Toronto and two other stations). Rogers Telecommunications was controlled by Ted Rogers.

The Jarmain family, which had pioneered cable in London, Ontario, held 7.7 per cent of the voting shares of CCL. The company's management other than the Jarmain family held 1.6 per cent. That gave the management block, which was the effective licensee responsible to the CRTC, 9.3 per cent in total. The major share blocks belonged to three parties — Jonlab Investments Ltd., controlled by Brascan, 25.8 per cent; Royal Trust (for clients) 18.3 per cent, and Montreal Trust (for clients) 9.4 per cent. These holding companies were passive shareholders.

Rogers had tried earlier, in 1974, to get control of CCL, but had been foiled by Jonlab, whose president and chief executive officer at the time was A. F. Griffiths. The same Griffiths, in 1977, was chairman of CCL. He had remained on the Jonlab board until the spring of 1977, a vantage point where he could keep any eye on the status of Jonlab's block of CCL shares. Just a few months after Griffiths left the Jonlab board, Rogers managed to pick up the Jonlab block, while Griffiths lost out in the shuffle. (The history of Rogers's relationship to the Jarmains and to Griffiths regarding this episode, and whether he acted in an open and obliging fashion as he claimed, or in a devious manner as they claimed, filled pages and pages of letters, memoranda and notes addressed to the filing cabinet, all of which were submitted to the CRTC hearing.) Then Edper Investments, controlled by Edward and Peter Bronfman, bought up the Montreal Trust block and shortly after would add other, widely held shares, for a 24.2 per cent interest. The commission in the interim asked that no further acquisitions take place pending a review. It then withdrew the request and announced that as far as it could tell a transfer of control of CCL had not taken place. Then it changed its mind and called a hearing (hearing number one). It turned out that Rogers and Edper had a buy-sell arrangement that could be triggered by either side, putting exactly 50 per cent in single control. But both Rogers and Edper denied any attempt to control CCL, defending their share purchase as an investment in the industry. The commission, however, decided that the stock dealings did constitute effective change of control of the company and that a takeover hearing had better be held. Rogers and Edper thereupon could not agree, or chose not to agree, on a joint submission for such a hearing, upon which Edper offered to buy Rogers's shares in CCL at a lucrative $18 per share (shares had been trading the previous year at $13.50). The buy-sell agreement allowed Rogers to match the offer and take over Edper's CCL holdings instead, which it promptly did. Edper presumably pocketed a nice capital gain for its troubles.

During all the manoeuvring, CCL management had refused to allow Rogers or Edper representation on the CCL board. They distrusted Rogers's motives. They saw him as trying to worm his way into control which, in the light of events, was not unrealistic. There was bad blood between Griffiths and Rogers, the former not liking the way the latter did business. CCL management looked down on what they considered Rogers's dictatorial management style and, huffily, on his spending a "lot of time doing things that aren't necessarily consistent with traditional management practices." "Ted's a people eater," said one industry insider. (On one occasion, in an abortive attempt to take over Premier Cablevision of Vancouver, Rogers had entered the Premier board room with a three-piece Hawaiian band and four hula dancers. "I wanted to show them the cable business was now show business," boomed Rogers. This was not exactly CCL's style, either.)

At hearing number one — I have numbered them to enable confused readers to keep track — called to sort out who owned what and had what arrangements with whom, the lawyer for CCL denounced Rogers's "predatory tactics" in the industry and accused him of distorting facts and misleading the CRTC to cover his tracks. "It was," said a newspaper account, "the most caustic personal attack on one major operator by another in the annals of the CRTC." It also happened when two of Rogers' small children were at the hearing, sitting in the front row.

"I knew we were going to get some flack," said Rogers later. "And I think it's well for children to grow up knowing the world is not always a nice place. It prepares them for life." It also prepared the CRTC to know what a family man Rogers was. The slow-witted Canadian Cablesystems management, who spent their time running cable systems, did not even think of taking their small children out of school and sitting them in the front row.

Andrew Roman, general counsel of the Public Interest Advocacy Centre, acting in this case for the Canadian Broadcasting League, remarked sarcastically, in a brief to the commission, that the CRTC was "like a schoolboy playing Pin the Tail on the Donkey," instead of a branch of government with the powers of a superior court of record. Trying to reconcile public regulatory responsibility with corporate games, where the players, moreover, were not even obliged to tell the CRTC what they were doing, was, Roman pointed out, trying to reconcile two mutually exclusive considerations. "All the King's horses and all the King's men — even those employed by the Commission — cannot reconcile them." He said it was unacceptable for a licensing authority to have to stand up in public and, "with some embarrassment, ask, 'Will the real licensee please stand up?' " He wondered what would happen if control was difficult to establish and nobody stood up.

The press slavishly reported the colour and the antics of the CCL-Rogers-Edper shenanigans and the details of the corporate transactions, as well as the CRTC's attempts to keep track. This gave the appearance that those happenings had legitimacy in determining who should apply and ultimately have a chance of getting the licences. Only Geoffrey Stevens, the Ottawa columnist of the *Globe and Mail,* had the wit and the independence of mind to get at what was really happening. He pointed out that the private stock dealings and other corporate manoeuvres should have had no bearing whatsoever on who had the right to apply for or to control cable licences. The commission had lost control.

Stevens wrote a classic two-part series, "Who's in charge?" and "The empire builders." He took up the issues one by one: how Rogers was attempting to get the CCL licences through the back door of a corporate takeover where it would not have been able to do so through the front door of open licence hearings; concentration of ownership by the manoeuvre; trafficking in cable licences; the logic of competitive hearings when operators wanted to sell out. Private meetings the previous fall between unidentified CRTC members and major CCL shareholders, which created an appearance of bias for future proceedings, had also become an issue. Despite Stevens's reputation and the status of the *Globe and Mail*'s Ottawa column, the articles had no discernible impact.

So hoary was the affair that before the actual takeover hearing (hearing number three) took place, the commission was obliged to hold a prehearing conference (hearing number two) to straighten out procedural complications. Nearly a dozen lawyers representing CCL, Rogers and Edper Investments argued about the hearing's format. The commission decided that the issue before it was "whether . . . RTL [Rogers] should have effective control . . ., and not whether the group which presently exercises effective control should retain that control." In other words, Rogers would be the sole applicant whereas CCL management, the effective licensee, though allowed speaking privileges of an applicant for the occasion, would be in effect only an intervenor!

CCL operations were widely considered in the industry to be superior to Rogers in key respects, particularly its decentralized management style and its technical work. CCL, for example, had twelve professional engineers on staff; Rogers had none. Now the CRTC had relegated CCL management to inferior status next to Rogers.

The collapse of regulatory order produced one other deliciously perverse circumstance. Rogers could not unilaterally apply for the CCL licences because they had not been surrendered. Rogers could not force CCL management, the effective licensee until otherwise decided, to apply for a transfer of control of the licences, either. Nor could the CRTC. But how could the CRTC hold a hearing to consider giving Rogers control of the licences when there was no application? They couldn't.

In a proper licensing situation this would not have been a problem. As Roman patiently pointed out, the holding of the licences by CCL management assumed they had functioning control of the company as well. This was a fundamental condition of the licences. When it no longer held, the licences became null and void, and new licences could be issued.

If new licences were being issued, however, somebody could argue that there should be competitive hearings open to any interested party — something the commission was earnestly avoiding. It may have been the only case in recorded history where a regulatory agency, making up its own rules, checkmated itself.

CCL management unfortunately had no sense of humour in this matter, and less sense of theatre. They themselves did not want anything so upsetting as licensing reform. So to expedite matters, like condemned victims obligingly digging their own graves, they took responsibility for filing the application. "I think we are all agreed," said CCL chairman Griffiths when the takeover hearing finally took place, "that the application is somewhat unusual, since the applicant is not in favour of the subject matter of the application."

Having managed to push the dithering CRTC into a takeover hearing by his stock-dealing putsch, Rogers began pouring cash into the "application." CCL fought back, after a fashion. By the time they were finished, what should have been a one-page dismissal of "coerced takeovers," based on a couple of short affidavits, had turned into a stack of twenty-three CRTC volumes thirty inches high. Rogers Telecommunications's part of the application alone was three-and-a-half CRTC volumes (five inches). The CCL filing made up two CRTC volumes, the first a seventy-seven-page bound book and the second a fourteen-part, one-inch-thick treatise. Rogers's reply, answering in minute detail — this was not the "application" or an intervention, but a mere rebuttal to an intervention — was three-quarters of an inch thick. Rogers's spoken presentation was produced in a one-half-inch bound volume. There were approximately one thousand interventions with CRTC acknowledgements, in fourteen volumes that stacked up seventeen inches high. Rogers responded to them all, then logged them, showing the name of the intervenor, the intervenor's address, what was considered the thrust of the intervenor's intervention, and the response to each. This appeared in what was called the Green Book. If the CRTC wanted to know what the interventions were about, they just had to look at Rogers's compendium and its handy conclusions, packaged for the commission's convenience.

For the purposes of the package, Rogers commissioned ten consultants' studies. Most of them were on policy subjects. Four of them were on size and concentration; two of those were American and a third by obliging broadcasting lawyer Jerry Grafstein. None of them dealt with the social and political implications of concentration, least of all in a Canadian context. That was not, of course, why they were commissioned. They made a hefty and impressive-looking volume.

Expert witnesses like consultants are, in proper administrative proceedings, as in legal proceedings, open to cross-examination, where their credentials, methodology and objectivity can be tested. This was the case in the CRTC's telephone-rate hearings. Given the assumptions of expertise of expert witnesses and also the scope and technical detail of their submissions, the right to cross-examine is a necessity. Since an intervenor does not have the time, and probably not the money, to commission and collect an equally massive counterweight of consultants' studies, cross-examination is also necessary to demystify the pretensions of the consultants' studies.

Rogers probably knew, however, that in a couple of cable rate-increase hearings, where the issue had come up, the CRTC had denied the right of cross-examination. It could be counted on to do the same in this cable hearing, despite the introduction of consultants' work. The commission had constructed a simple-minded myth about informality in broadcasting hearings, in contrast to the complications of telephone hearings. The idea was to ensure that the ordinary citizen could participate in comfort, without lawyers and without sophisticated skills. When the applicants introduced complications, however, or when the issue, like rate

hearings or concentration, was inherently technical or complicated, and where the applicants themselves had recruited platoons
of lawyers and consultants, the myth became a dead letter. On
the other hand, the commissioners would likely hang on to the
myth for dear life. If they allowed cross-examination at broadcasting and cable hearings, they would have to sit even longer, perhaps for extra days. This is the true and heretofore unexposed
explanation of the cross-examination question: the buttock factor.
The other solution was to exclude the consultants' studies from
the application, which Rogers could anticipate the commissioners
would not do. The studies were included, and the right to cross-
examine was duly denied.

Next came the old policy-ploy-in-the-shape-of-a-licence-hearing
trick. Section 1 of the Rogers Telecommunications package was
called "Benefits to the Canadian Public." The first section of that
section, after the table of contents, was "AN OVERVIEW — CANADA
AND CABLE TELEVISION," in which Rogers, in bold patches, pronounced upon broadcasting and cable policy.

> These are uncertain and changing times in the country, and the broad-
> casting system reflects these times. . . . Canadians are buffeted by pres-
> sures of an attractive and pervasive cultural message from the south.
> They are aware of the fragile nature of the English and French partner-
> ship; and are also aware of the pressures for recognition of the scores of
> other cultures that make up Canada. Our nation is in trouble, a nation
> divided, divided in many ways. But it is also a nation that is capable of
> coming together.

However, all was not lost. "LET'S MOVE NOW — MORE CANADIAN
VIEWING" announced the submission. Rogers was committed to "the
repatriation of viewers to Canadian-made programming," and
would show how it would save the day.

Names and quotes were thrown in to suit in this and subsequent
sections. Pierre Juneau was there, Harry Boyle, Al Johnson, Minister of Communications Jeanne Sauvé. There were quotations from
the CBC's "eloquent and magnificent renewal application" about the
idea that broadcasting must concentrate on nation-building. Select
quotes from the consultants' studies were also pasted in. The studies did not necessarily represent Rogers's position or support it, but
appropriate citations can always be found if the studies are long
enough, in the same way that a theatre company can find a few
useful advertising words in most critics' notices.

But why were policy matters, like concentration, for example, being discussed at a licence hearing, where Rogers, a self-interested party, had the privileged right (1) to provide the primary documentation; (2) to commission the research documents according to its own choices and priorities, and (3) to have this material filed for public examination before the hearing? Rogers, in this quasi-policy hearing, also had the special rights of an applicant, including extended presentation time and right of rebuttal. Further, these central policy questions were being dealt with at a local hearing, in an Ontario city, for which the notice had been posted only in those places whose cable licences were directly at stake. In the same spirit, the official notice of the hearing was marked only "Ontario Region," although there was also a Calgary licence involved.

The matter of concentration and other related policy questions, if dealt with by the CRTC at all, should have been assigned to a national policy hearing, where the previously filed policy papers and research documents, if any, were prepared by the CRTC, and where all parties had equal procedural rights. Properly, a question like concentration was a matter for Parliament.

The bait for the commission was a string of promises, a "shower of goodies," as one newspaper account frankly put it, a bit of a sugar-coating to help slip the takeover through. When one took a close look, most of these "goodies" turned out to be marginal. There was, for example, a "Something for Something" fund, to which Rogers committed 1 per cent of gross revenues, to be used at the outset for a repeat channel. The outlay was minor, and the cost would eventually be passed on to subscribers, but the name was sure catchy and, under the heading, the applicant could make plenty of devout affirmations.

There was a "social dividend," or "social benefit," in the same vein. One part of the "social dividend" was the development of a French for Canadians Channel, a three-hours-a-night repackaging of existing material at a throwaway cost ($225,000 a year), to which Rogers was committed for only two years. The gesture could nevertheless be well appreciated by the commission's heavy Quebec contingent. Another of the goodies was an offer to aid cable-licence boundary rationalization in Metropolitan Toronto by transferring out 65,000 subscribers — not such a bad trade for the 442,000 subscribers Rogers would get in exchange. A regulatory agency should have looked after this matter independently, but the CRTC did not have even the minimal regulatory initiative to do so. Another goodie was a research-and-development commitment, something too that should have been integral to the commission's overall regulation of cable. Filler on other subjects was added, listed complete with descriptions sufficient to make this section of the document respectably thick and to demonstrate that little bit of extra enthusiasm.

The Rogers application was grandly transparent. In the regulatory disorder, nobody blinked. No competing applications were allowed, of course, to show what really could be done with the cash-generating capacity of CCL. That was the beauty of "coerced takeovers."

The gathered multitude at the Auberge de la Chaudière in Hull, where the takeover hearing was held, resembled the cast of a small Cecil B. DeMille spectacle. CCL management fielded thirteen people for its presentation. Rogers fielded eleven, plus three from Edper. Each of the two parties had two and a half hours, rather than the usual twenty minutes. Rogers came up with "expert witness" André Raynauld, former head of the Economic Council of Canada and a good Quebec federal insider. CCL produced Douglas Hartle, distinguished professor of economics at the University of Toronto with impressive government work in his curriculum vitae. The joke went around at the hearing that so many broadcasting lawyers had been hired by the two sides collectively, some put on retainer to keep them away from the other side, that there weren't any left.

Any innocent who walked into this corner of the licensing circus could only end in desperate confusion. One Hamilton contractor who did cable work for CCL and was partial to their management decided to intervene in order to bring his independent common sense to bear on the proceedings. He got more than he bargained for. "My opinion," he told the commission after vainly trying to figure everything out, "is that it is a 'dog-eat-dog' story, and whoever misleads the other more and gets away with it, ends up on top."

Poor CCL management! They argued that CCL, not Rogers, was the leader in technology; that the consultants' studies on concentration were hopelessly vague; that any "social dividend" from the use of the licences was possible regardless of who the licensee was; that CCL plans for the licences were superior to Rogers's; that any alleged economies of scale could be achieved through co-operative mechanisms without the need for merger; that the suggestion that the commission was faced with a *fait accompli,* implicit in the Rogers submission, was incorrect and indefensible.

These were all splendid rational arguments. But by allowing Rogers to be an "applicant" in the first place, and the only applicant, instead of asserting its licensing authority against the stock-dealing manoeuvre, the CRTC had already gone halfway to giving Rogers the licences. Further, for a commission that had detached itself from the footing of basic public administration principles, rational reasons were no match for the Rogers's four-volume, five-and-a-half-inch package, its quotes and names, all those consultants' studies (whatever they said), the protestations of cultural nationalism, the Green Book, and the rest. Like a mud slide through a fence, the merger went through.

The commission's decision extolled the formation of "an economically more powerful unit." It did not, however, provide details to show why a combined Rogers-CCL operation of 572,000 subscribers was so much more economically advantageous than a CCL operation of 442,000 subscribers and a Rogers operation of 195,000 subscribers working in tandem, for the very good reason that there were no such details. Economies of scale in cable are basically local, and what isn't can be developed easily on a co-operative basis since cable licensees, each having a monopoly, do not compete against each other.

The affair had so much curious detail that nobody, including Capital Cable Co-operative, noticed that the takeover proceeding amounted to a competitive hearing at renewal time, albeit one of a peculiar sort. A licence challenge had been made and was duly heard. The hearing took place regardless of the preferences and timing of the existing licensee. The promises of the challenger were carefully listened to and compared to the past performance of the existing licensee. This latter point was spelled out: "The level of service . . . by the [Canadian Cablesystems] undertakings and the benefits . . . generally . . . are clearly relevant as standards against which the [Rogers] proposals are to be measured."

Then, at the November procedural hearing two months later, the commission tried keeping up the pretence that a comparative procedure could not be done — had too many difficulties and drawbacks — while, for purposes of the Rogers takeover, it had just done it — was right in the middle of doing it. Unlike the takeover affair, a proper competitive procedure would even have let the incumbent licensee try for the licence, too.

7

The Demise of
Local Ownership

THE MUCH-INTONED, hallowed idea of local ownership is another entry in our story. In a speech in Winnipeg in 1973 Pierre Juneau proudly explained "why our whole broadcasting system is so decentralized with literally hundreds of locally licensed small, medium and large radio, television and cable operations in all parts of the country" and "why the CRTC prefers local ownership of radio, television and cable outlets whenever this objective is feasible." There was Toronto journalist Jack Batten's breathless conclusion, in that 1972 account, that "*small* and *local* took precedence over *big*" in the CRTC's wise and courageous thinking.

In 1976 Capital Cable Co-operative wrote to CRTC chairman Harry Boyle calling for the divestiture of multiple holdings. A planning official, responding for Boyle, wrote that "the commission does consider local ownership an important principle, and in fact undertakes the assessment of competing applications for cable television licences by considering this principle among others as desirable and primary." Decentralized ownership and control of private stations and cable systems was a clearly enunciated value. As late as July 1978, in the working paper on procedure, the commission pronounced that "if a proposed transfer involves the replacement of local ownership and control with non-local ownership and control, the applicants must satisfy the commission that there are no prospective satisfactory purchasers or that the proposed transaction is otherwise particularly in the public interest."

At the same time, concentration of ownership kept increasing pell-mell. I have already mentioned how, in the Juneau years, Standard Broadcasting (Argus Corporation) took over CJOH Ottawa; a Bronfman family subsidiary picked up CFCF Montreal; Selkirk

Holdings took over CHCH Hamilton; Premier Cablevision, already holding multiple licences, was given the Oakville and Keeble Cable (Toronto) licences; Agra Industries picked up six different cable licences in smaller centres; Canadian Cablesystems and Maclean-Hunter Cable each acquired another cable system; Baton Broadcasting (Bassett and Eaton families, CFTO) added four radio stations and the Saskatoon CTV station; CHUM Limited (Alan Waters, with controlling interests in six radio stations, picked up CKVR-TV Barrie, CJCH-TV Halifax, CJCB-TV Sydney, CKCW-TV Moncton, CKVN (call letters changed to CFUN) radio Vancouver, CFRW-AM and CFRW-FM (now CHIQ-FM) Winnipeg.

This pattern continued. Agra Industries picked up the Courtenay-Comox cable licence and a small Toronto cable system (Wired City); Selkirk Holdings got radio stations in Toronto and Sarnia; the principals of Scarboro Cable Communications took over the cable licences for Barrie, Orillia and part of Mississauga; Maclean-Hunter acquired CKOY and CKBY-FM Ottawa and CHNS and CHFX-FM Halifax; Multiple Access (CFCF) and then CHUM Limited picked up City-TV; Canadian Cablesystems acquired one of the two Calgary cable licensees and then was taken over entirely by Rogers Telecommunications, and Premier Cablevision acquired 45 per cent of Western Cablevision (Surrey, New Westminster, Abbotsford and Clearbrook). These takeovers took place up to and including the summer of 1979, and they are only the major approved takeovers involving concentration. There were a few other attempts that were denied by the CRTC or that were abandoned, including several for Premier Cablevision (of these, two were by Western Broadcasting) and Baton Broadcasting's run at CFCF Montreal. These denials or abandonments only meant that other takeover applications would soon surface as new trafficking deals were arranged.

How could the accelerating concentration occur in the face of the local-ownership policy? It was not in the least a mystery. To start with, the concentration had nothing to do with merit, inevitability, competence, the advance of technology, superior tactical intelligence, or anything like them. The key was the exclusion of competing applications, and the licence-trafficking that went with it, under the protection of the CRTC.

A retiring licensee trafficking in the licence would go for the highest bidder. Because of existing assets and cash flow, an existing company in the industry already had a line of credit to set up deals for further acquisitions. The large multiple-function or multiple-licence communications corporation also had the cheapest and most available credit, up to an organization like Maclean-Hunter, which could float a debenture at a rate that no local company could approximate. The cheaper the credit to finance the acquisition, the higher the possible offer. In addition, the larger and the more diverse the aspiring takeover corporation was, the more manoeuvrability the corporation had, through intracorporate dealings, in writing off the cost of the acquisition against taxes — that is, in getting the Canadian treasury, and hence the Canadian taxpayer, to help pay for the acquisition. Hence also the more lucrative the offer it could make.

Where multiple licences were involved, a company or society could not apply only for the licence in its own locality or even enter the trafficking auction for it. The deal usually took place for the package. The integrity of the individual licence had long since been abandoned by the commission. *The individual licence, on which the whole licensing system was supposed to be based, did not come into it.* The larger the package was, the larger the takeover company had to be. Even in a reverse takeover — a smaller holding company taking over a larger one, as Rogers had done with Canadian Cablesystems — the object was to capture the package, not to extract one of the licences and surrender the rest.

The takeover company had then to get its deal past the CRTC. Here again the large company had an advantage. Because of its activities, it had a presence in society and an assumed credibility based on its commercial weight alone. It was already a licence holder and also part of the intimate web of relationships between the CRTC and the industry. And it had the resources and people to maintain contacts and to lobby.

Sooner or later takeovers would get through. Simply by dealing with takeover applications by organizations that already had licences, the CRTC permitted concentration. This message was the hidden meaning of official descriptions of the process, like the following, from public notices (the italics are mine): "[The commission] has . . . dealt with ownership applications on the merits of each individual case, *taking into account the particular factual circumstances.*" And: "It is proposed that these [policy requirements regarding licence reallocations, or so-called transfers] remain as matters of policy only, and not law, as they must be applied with discretion and understanding, *taking into account the particular circumstances of each case.*"

"Taking into account the particular factual circumstances" sounded impressive. The trouble was that a whole series of individual takeover cases represented more than "particular factual circumstances." It represented a structural shift. Without a lucid and well elaborated general policy behind local ownership, the commission's "individual case" approach inevitably led to greater concentration, despite stated policy. Any schoolchild could have anticipated it.

But the excuse of "particular factual circumstances" was phony to begin with. The whole takeover scam was greased by the exclusion of competing applicants. As in the Victoria cable case, the "particular factual circumstances" might indicate that local control of the licence, rather than conglomerate control, would be of enormous benefit to subscribers and to the community and to the objectives of the Broadcasting Act. But the local applicant never got a look-in. Consideration of the principle of local ownership as "desirable and primary" in the "assessment of competing applications" — the answer in Harry Boyle's day to Capital Cable Co-operative — when competing applications were not allowed, was a vacuous piety.

What, too, about the "particular financial circumstances" in connection with divestiture and deconcentration? Aside from general policy, a good case could be made in individual instances for divestiture and breaking up of multiple holdings — for local control of the cable licences held by Premier Cablevision, for example. "Particular factual circumstances" were never talked about in that way. Individual cases were decided not on "particular factual circumstances," but on fictional circumstances, in which the commission made believe that options did not exist. There might be stop signs on the road to concentration, maybe even roadblocks, even from time to time what appeared to be barricades, as in Pierre Juneau's cosmetic declamation on local ownership, but the street was quite clearly marked "One-Way," and there were no cross streets. No wonder all the traffic ended up at one end.

But why did the commission not establish a lucid policy of local ownership, with a full elaboration of its political and social principles, and with machinery in place for interlicensee co-operation where useful? Well, the commission, going back to its initial days, just did not get around to it.

Gradually, as concentration increased, a rationale had to be found for it. The commission accordingly produced the "significant benefits" rule. Where concentration of magnitude was proposed, said the rule, "there must be significant and unequivocal benefits demonstrated to advance the public interest." This was the "particular factual circumstances" excuse tarted up. It left plenty of room for the CRTC's disposition of the moment.

In most cases the takeover company could cook up something of a package to qualify under the "significant benefits" rule without much extra financial outlay, and well worth it for the acquisition payoff. Further concentration followed apace. Still left out of the rationale, of course, was the possibility that local applicants, were they allowed to apply, might provide even more significant benefits, indeed might provide major benefits that would put the exclusive takeover application in the shadows.

The "significant benefits" falderal had another wonderful twist. The worse the licensee was — the more it dragged its feet, held back on investments, ducked promises and ignored CRTC orders — the easier it was for a takeover company to promise "significant benefits," not necessarily exceeding what the outgoing licensee should have been doing in the first place. So the reprobate licensee, instead of being punished for poor performance by a denial of the takeover application, was rewarded by the ease with which the take-over went through and by the high price it could command. Allow-ing the takeover and the attendant concentration was a way the commission had of getting the operation to do what the commis-sion, as regulatory agency, was supposed to have compelled the licensee to do. It was a way, in short, of making up for the commis-sion's own failure. Removing a reminder of the commission's embarrassing failure was a heaven-sent psychological lever for get-ting a takeover through. The commission could alternatively stick with the undesirable licensee and punish it by denying renewal. But refusing to renew a licence was, in the commission's repertory, not done.

The existing licensee would exploit this situation for as long as it had the stomach for it. Then, at a judicious moment, it would flog the public licence to a takeover company, adding to concentration but with the commission's co-operation and to the commission's great relief. One classic example in the early 1980s was the 12,000-subscriber White Rock, B.C., cable system. White Rock Cablevision had dragged its feet shamelessly on community programming, had alienated its employees, and had not modernized enough to carry pay-TV. Its licence should have been lifted for cause. It then arranged a closed takeover by Capital Cable TV Ltd., an Edmonton multiple-licence company not to be confused with Capital Cable Co-operative. Of a transaction price of $3,150,000, only $555,000 represented shareholders' equity. The rest, $2,595,000, was the trafficking margin from the value of the public licence, a ratio to equity of almost 5 to 1. The original investment, which was also the total share capital to the date of sellout, was a nominal $13. Not a bad payoff, at the public's expense, for a party that should have had its licence revoked. The outgoing licensee, in the reallocation application, tried at first to keep the transaction price out of the public file, on the grounds that if the public knew about it, he or other members of his family might be harrassed or kidnapped for ransom. What actually happened was that he ransomed off the public's licence. The CRTC was only too glad to help him out and finally get rid of him. No competing applications, including applications from the White Rock area, were allowed.

The main function of concentration in all this — I disclose it here for the first time — was as a therapeutic device for owners and financial management. Acquisitions helped to keep them occupied and to justify and augment their status and self-image in the scheme of things. Arguments about economies of scale and benefits would have to be cooked up to suit.

The CRTC, in one of its own notices in 1979, hinted at the situation: "With the near completion of the extension of cable television services to most population centres of significant size, and an increasingly crowded on-air spectrum, present opportunities for new licences are limited and a major current initiative in private broadcasting is for growth by acquisition or merger."

To add to the pressure, potential new services in cable, like pay-TV, which might increase the stream of earnings, were bogged down at the time by regulatory obstacles, most of all because schemes like pay-TV did not make Canadian sense. The *Financial Post* put the problem most clinically, delightfully. In one of its annual issues on the mass media, it reported how the larger cable companies were "working off their expansive energies on the stock exchange and in the boardroom," and how expansion through acquisition was "one of the few relief valves available."

These unsublimated board room energies just had to be accommodated, didn't they?

8

The Very Naughty Book

IN THE SUMMER OF 1979, shortly after the working-paper fiasco, the Economic Council of Canada published *Canadian Television Broadcasting Structure, Performance and Regulation,* by economist Robert Babe, a communications specialist. Babe began his work with the statutory mandate of the CRTC — to regulate and supervise the Canadian broadcasting system so as to ensure that broadcasting will "safeguard, enrich, and strengthen the cultural, political, social and economic fabric of Canada." In a cautiously worded, academic, methodical manner, and using case studies where appropriate, he proceeded to examine how well the different parts of the structure were serving that mandate and how the CRTC handled major regulatory and policy matters. Each section was heavily footnoted and a lengthy bibliography was provided.

The effect was telling because the facts were telling. Babe documented how the CRTC had allowed the de facto establishment of private property rights in the public radio spectrum through trading in the profitability of licences, automatic licence renewals, weak enforcement of original licence conditions and the exclusion of competing applications. He found the CRTC's excuses for excluding competing applications irrelevant and recommended competitive hearings. The only obstacle to competitive hearings, he observed, was the CRTC's lack of will. Babe also questioned the CRTC's paternalism towards private licensees. This "personalization" of regulation avoided the basic reality that broadcasting and cable had become big business and that the dominant imperative of big business was financial. These financial interests were diametrically opposed to the lofty goals of broadcasting. Broadcasting in Canada was "no longer, if it ever was, comprised principally of small-time businesses run as proprietorships and as a public service to help foster local expression and national unity." The problem, Babe pointed out, was that the CRTC tended to view the large diversified companies engaged in broadcasting and cable as public-service or charitable organizations, rather than as corporations with outlooks that are substantially similar to those of corporations in other areas. These corporations valued licences in terms of the stream of earnings they were expected to generate therefrom. Babe's book cut through the CRTC's fuzziness and fakery by dealing with cases, facts and figures and sticking to basic premises of public administration. Here and there, newspaper and magazine articles had taken pokes at the amazing behaviour of the CRTC — Geoffrey Stevens's items in the *Globe and Mail,* columns of my own on the editorial page of the Toronto *Star.* Babe's book was the first analytic treatment between covers, however, and it had the Ottawa imprimatur of Economic Council publication, to boot.

While the study was still in preparation, the CRTC began playing bureaucratic games with the Economic Council over its publication. The CRTC's executive director, Michael Shoemaker, in a letter to the vice-chairman of the council, took the liberty of suggesting that "if for any reason the Council decided not to publish Dr. Babe's work but was thinking of granting permission to have it published independently . . . you would seek the advice of the commission as to whether permission should be granted." Shoemaker later, on reading a draft, huffed to the Economic Council that

the whole study has a tone of hostility towards the CRTC that might make good reading for many but which seems to diminish its quality as a serious study of many of the complex problems of regulation of the Canadian broadcasting system. Without in any way attempting to abridge Dr. Babe's right to hold his opinions and to express them publicly, it might be questioned whether the Economic Council of Canada should appear to endorse these views without a disclaimer which would indicate that the views were those of the author.

A month after publication, CRTC chairman Pierre Camu wrote to council chairman Sylvia Ostry that "this study is considered to be an exercise in polemics." Accusing somebody of being polemical is equivalent, in Ottawa bureaucratic circles, to accusing them of bad manners and body odour at the same time. "Should aspects of his paper result in serious questions being raised in the House of Commons . . . or in the press," Camu blustered on, "the commission might find itself in the position of having to make [an earlier critical letter to the Economic Council] public, as one alternative."

The Canadian Association of Broadcasters, meanwhile, charged that Babe detested private enterprise. His study could not be treated as a work of serious scholarship, they claimed in a letter to Ostry, because it was full of Professor Babe's personal opinions.

What a lot of huffing and puffing. Yet Babe's book had in fact told just a part of the story.

Camu, the man who was going to make the CRTC the best-run regulatory agency in the world, judiciously left shortly afterward, with the Clark government now in power. His departure was not loudly lamented.

9

The Fall of British Columbia

THE AMAZING SUCCESS of the putsch of Canadian Cablesystems Ltd. (CCL) by Rogers Telecommunications had shown just how enfeebled the CRTC actually was and how a privileged takeover could be finessed if the right techniques were used and enough money was spent on them. A whole shelf of public cable licences had been snaffled via a stock-market manoeuvre, also against the wishes of the existing licensee (a factor the commission previously considered positively sinful). Similarly, Rogers's casual discarding of the declaration that he was not interested in taking over CCL, a declaration behind which the stock-purchase deal and the buy-sell arrangement with the Bronfmans had initially been passed off, had resulted in no ill effects at all.

The angry charges that Rogers was a manipulator and not fit to get the licences only put the accusers in the shadow for being so personal and unseemly. The best part of all, the real key to success in gaining the crucial power of licence control, was the absence of competing applications. No takeover syndicate could ask for, or could hope for, more acquiescence from the licensing agency, for future gambits.

The earnest and ambitious Rogers, son of the inventor of the batteryless radio, had for some time coveted the licences held by Premier Cablevision of Vancouver. Efforts to take over Premier in 1976 and 1977 had aborted before coming to hearing. The idea of a takeover of that size, and increased concentration, was brash even then. Now Rogers had swallowed Canadian Cablesystems. He had tripled his size and moved from the fifth-largest cable company in Canada to the largest, with 648,000 subscribers by August 1979, outstripping Premier by 133,000 subscribers and achieving a concentration until then considered unallowable. He now was going after Premier as well. When a second attempt by Western Broadcasting to take over Premier failed to get by the CRTC, Rogers moved in. In November 1979 he made an offer to Premier's controlling shareholders, the original management group, that they would not want to refuse.

The takeover, if approved by the CRTC, would give Rogers 1,112,000 subscribers, 27 per cent of the country's total. It would give Rogers B.C. cable licences representing 51 per cent of the provincial total, including the key licences in Vancouver and Victoria. The licences for Coquitlam, B.C., and Keeble Cable and York Cablevision in Ontario also came with the package, plus 45 per cent of Western Cablevision (Surrey, New Westminster, Abbotsford).

The takeover would wipe out major British Columbia entrepreneurship in cable and the licences' exceptional potential if brought under subscriber ownership. It would do so by surrendering the most advanced subscriber participation in the country, in Vancouver and Victoria, to a personally held Toronto corporation. No British Columbia organizations or companies were allowed to apply for these British Columbia licences.

The licences, moreover, were the key around which the financing turned. Whoever had the licences, with their established revenue in this case, could also arrange the financing to acquire the systems and, through the licences, then gradually retire the financing. Here was a case of a personally controlled Toronto corporation given monopoly licences whereby B.C. cable subscribers would help pay for the acquisition of their cable systems by that corporation while, at the same time, they were forbidden to apply so that their payments could go to acquiring the systems under their own ownership.

The mythical notion of an important moneyed party qualifying as a potential licensee because of the investment it was prepared to make through the boldness of its entrepreneurial heart, had the scenario backward.

By the beginning of 1980 the CRTC had a new chairman, John Meisel, a political science professor at Queen's University. Meisel had been appointed by the new Conservative prime minister Joe Clark and Minister of Communications David MacDonald. Flora MacDonald was also involved in the appointment. Allan Fotheringham, for one, was enthusiatic:

> The fact that the prime minister ... has the wit and courage to name someone such as Meisel as Canada's new broadcasting czar gives one

small tiny glow on the first weekend of Ottawa's frigid winter. . . . That it's a good appointment is obvious — once you know Meisel's background. . . . His views on the terrible state of the pseudo-Canadian TV industry are well known. He has said that the country has sold out "the dreams of its forefathers" by allowing commercial broadcasters to turn our TV into a brainwashing instrument, with its reliance on American pablum for the mind.

He complained, in a 1977 paper, of "the highly developed commercialism of Canadian values." He said "Canada's toleration for foreign control and influence over its economy, and culture, so strikingly visible in its school curricula, its television programs and practically everything we do, is unique. No country has so supinely encouraged the takeover of its resources of mind as has Canada." . . .

The appointment, which Clark had said was the most high mandarin decision he had to make, is so welcome, considering some of the tired minds of commercial broadcasting who were suggested.

When Meisel, the "passionate nationalist" as he was described at the time, first refused the offer, Flora MacDonald told him he just had to accept. "The country needs you in this job," Prime Minister Joe Clark told Meisel.

"I had respect, contentment, minimum pressure," Meisel explained to a journalist. The calls to duty overcame his objections.

The cable industry's reaction to the Meisel appointment was fierce. Private station owners were also very upset. Meisel responded with conviction. "Everybody likes to be liked . . .," he said. "But there are certain principles the CRTC has to follow. It has no choice — they're in the law. And if that means getting some people mad at us, then the principles will have to come first. We can't back off."

Shortly after Meisel's appointment, he talked about "inducing" commercial broadcasters to commit themselves to Canadian programming objectives, despite the interviewer's reminder that just that strategy had failed so badly for the previous twelve years. It looked like business as usual.

Two weeks into Meisel's term, at the beginning of 1980, Capital Cable Co-operative wrote to the new chairman, alerting him to the takeover development involving the Victoria licence and to the past malpractice of the commission in excluding competing applications. The letter also catalogued the documentation. Its purpose was to ensure that Meisel had a grip on the necessary background independently of his bureaucracy. The crucial juncture, as Capital Cable knew, would not be the hearing on the takeover application, *but the decision as to whether the hearing, excluding competing applications, should go ahead as set up.*

The official request by Capital Cable to be allowed to apply in competition with CCL, Ted Rogers's new corporate vehicle, for any reissuance (so-called transfer) of the Victoria cable licence was duly filed in February 1980, well in advance of a public notice of the expected CCL application. This request was followed by a second letter to Meisel later that month, bringing to his attention the fact of the filing. This second letter chronologically detailed the commission's evasions, going back to 1969. It spelled out the critical importance of now finally stopping the bureaucratic foot-dragging, and taking action. Almost a year and a half had gone by since the 1978 hearing on procedure and four years since the Association for Public Broadcasting in British Columbia and Capital Cable had put the prior issue of competitive hearings to the commission.

Meisel, naive and slow on the draw, had commented in an intervening note that the issue "will, no doubt, be discussed at a forthcoming public hearing to hear the application of Canadian Cablesystems." The letter from Capital Cable patiently pointed out how this comment misunderstood and evaded the issue which, again, was not the merits of any CCL application but the "prior, fundamental question" of who should be allowed to apply for the licence in the first place. "Once into the hearing, that critical prior issue, for all practical purposes, has been bypassed." All the arguments and necessary information on the matter had already been heard many times, the letter added. What was required was some regulatory backbone.

The white-and-woolly-haired, casual, warm and amusing Meisel (as feature articles described him) ducked. The commission simply repeated its Orwellian fiction that no reissuance or even transfer of the licence was taking place because Victoria Cablevision Ltd., the licensee today, would also be the licensee tomorrow if the takeover were approved, although every man, woman and his or her dog knew that a takeover of the licence was involved, which was why the hearing was being held.

Meanwhile, Meisel was telling his clientele, at a conference of the Canadian Association of Broadcasters, that the CRTC "must be the articulators of the national conscience — the moral gyroscopes." Shortly after, he told a reporter that he saw the CRTC as a "moral force." He was learning to make the right speeches, on schedule.

"The CRTC . . .," Meisel told the conference, "seeks to co-operate with its clients." Blaik Kirby, television critic for the *Globe and Mail*, who did know something about the situation, wondered, in a feature on Meisel, whether he was "as much a pushover as, on the surface, he seems? Is he Superman," he asked, "or just Clark Kent?" Meisel had been on the job exactly four months. By this time Capital Cable Co-operative already knew the answer.

The payoff in the deal between Canadian Cablesystems (Rogers) and Premier Cablevision was enormous. What CCL's chief financial officer euphemistically described as a "purchase discrepancy" amounted to a staggering $61.6 million, the figure the company itself provided at the hearing, or two-thirds of the entire selling price of $87 million. This was for the value of licences that were not and never had been the property of Premier.

The speculative value of the Premier shares on the Toronto Stock Exchange before the deal was struck was $11.50 per share. In late 1976 Rogers had offered $15 a share, and it had been accepted. Rogers was now offering $25 per share in order to snaffle Premier for himself and hence be the exclusive applicant for the licences (the deal being provisional on CRTC approval).

In the case of the Canadian Cablesystems takeover of Premier, the price was so inflated that it could not be covered by the returns of the Premier licences alone, excessive though they were. The cash flow generated by the licences already held by CCL would help. The biggest help, though, was the ordinary Canadian taxpayer making up for lost corporate taxes. The growing Rogers empire would be able to make maximum use of tax breaks in the takeover. Maybe the person most admired by industry supporters of the takeover was the financial officer who had strung out the tax avoidance possibilities so well, reducing the *posttax* net cost of carrying the debt burden of the takeover, for example, from $16 million to $3 million over the first five years, according to the officer's proud testimony at the takeover hearing. "Fortunately, Revenue Canada is going to help us out," he explained, and, just to make sure his cleverness was not misunderstood by simple minds on the commission: "This is very legitimate."

The price had been established according to how far tax-avoidance provisions could be stretched. The proceeds went from the pockets of Canadian taxpayers into the pockets of Premier shareholders. The licences went into the pocket of the Toronto takeover company.

"It is a matter of record," the CCL takeover application read off the top, "that the controlling shareholder of Premier, Mr. S. W. Welsh, has been anxious to dispose of his interest in Premier for some time. Mr. Welsh has been unsuccessful in finding a buyer acceptable to the commission ... despite three previous attempts. ... Mr. Welsh [therefore] made contact with Mr. Rogers." The last chance!

This was a whopper of extraordinary proportion. For a start, Capital Cable had earlier made a formal offer for the Victoria system and been arbitrarily turned down by Premier. The offer had been equivalent on a per-subscriber basis to a deal Premier had made at the time with Western Broadcasting. At a pre-established level of compensation, in a competitive licensing procedure where anybody who could meet that level could qualify as an acceptable applicant — which is how the reallocation of the licences should have taken place — there would have been umpteen buyers of Premier's cable assets, as everybody knew.

In fact, under the closed family system, only one potential buyer for Premier's systems had previously been allowed to plead acceptability before the commission, Western Broadcasting, with two similar applications. As the president of Premier once frankly put it to the *Financial Post,* "We've turned down offers in the past year because the price was wrong, not because of the regulatory climate [viz. not because the potential buyer would be unacceptable to the CRTC]." By right price, as the *Financial Post* explained, he meant a big company which could pay big money for expansion [viz. that would pay a maximum price for the licences].

Premier's controlling shareholders had selected Rogers for the well-understood mercenary reason that his holdings were big enough for tax-avoidance purposes and that his ambition was demanding enough that he would come through with the high price. He might slip by the commission as well.

The $62-million margin for Premier shareholders, moreover, was on top of the high rates of return that Premier had been taking out of its licensee subsidiaries, rates of return not diminished by time. In the previous financial year, 1979, for example, the return on shareholders' equity for the Victoria system was an astonishing 68 per cent before tax and 33 per cent after tax. (For B.C. Telephone, also regulated by the CRTC, the equivalent return that year was 19 per cent before tax, 10 per cent after tax.) The return on the large Vancouver system, highly levered by debt, was 184 per cent before tax, 89 per cent after tax! Meanwhile, Welsh and the other remaining Premier original, vice-president Garth Pither, had already begun cashing in by selling to Premier their interest in several other companies, in two of which cases the Premier investment subsequently had to be written off and, in a third, where the company later ceased operation. For the CRTC, on the other hand, Welsh played the role of unselfish pioneer, to great effect.

Underneath its form and the necessary ritual of documents and questioning, the takeover hearing was a ceremonial gifting of the value of the public licences by the CRTC to Welsh and Pither, who were selling Premier. Western Broadcasting was also a major Premier shareholder and stood to gain. The routine came complete with a eulogy, following Premier's presentation, by novice Meisel, acting momentarily like a master of ceremonies.

Premier had nevertheless become a bit of an embarrassment even within the industry because of its unvarnished mercenary simplicity, which could not be hidden by hiring lawyers as leading officers. Unlike some of the smarter operators like Rogers, closer to Ottawa, Premier did not make the right sounds about Canadian objectives. Except for simulcasting, which did not cost much and was virtually impossible to be against, Premier had traditionally and openly opposed Canadian-minded CRTC proposals. It had not shown any enthusiasm for community programming, either. The commission had even felt obliged to take Premier to task for inadequate spending in Vancouver. Premier as well had been dragging its feet on modernizing its Vancouver system.

At the takeover hearing Premier's new president, George Fierheller, a Torontonian by way of Ottawa and a former fraternity brother of Rogers, freely admitted that Premier financially could do the various things that it should be doing and that Rogers was grandiloquently promising. It was not a matter of credit or retained earnings that was holding Premier back, said Fierheller. Premier, however, had no plans to undertake the mooted improvements with the same dispatch or at all. This cocking a snook in the face of the feeble regulatory agency, and the attempt to force its compliance to the takeover, was done unexceptionally and pleasantly, as if it were a perfectly normal and acceptable procedure. The agency, having long lost sight of basic principles of public administration, also took it as normal and acceptable.

Everybody, on all sides, wanted to get rid of neanderthal Premier. But without competing applications, the only way for the CRTC to get rid of it was to approve one of Premier's rigged arrangements. It was another pretty case, this time a huge one, of a selfish and recalcitrant licensee, instead of being punished for its inadequate performance, being rewarded by having a deal pushed through, just to get rid of it.

The real work of pushing the takeover through, however, was done by Ted Rogers, his two lieutenants Phil Lind and Colin Watson, financial officer Robert Francis (who presumably had worked out the dollar permutations and combinations) and assorted hired hands. Lind and Watson, two relatively young men with the right looks, tagging along in the Rogers cortege, were known as the Bobbsey Twins, or to the more jaundiced as the Golddust Twins.

The application Rogers had filed in late 1976 for the takeover of Premier, later withdrawn, had some bulk and pretension, but was nondescript. The 1980 application was basically no different in substance — a large Ontario cable company taking over a British Columbia one — except that the concentration involved, with the acquisition in the interim of Canadian Cablesystems, was much, much greater. But Rogers and Lind had, with practice, improved their packaging skills in the meantime. They had also realized the value of spending lots of money, conveniently collected from subscribers under monopoly licence, on the packaging and the lobbying. The earlier Rogers takeover of CCL served as a tactical proving ground for this one, with added refinements and with more resources now poured in. The strategic premise, found valid the last time, was that with an intellectually dim and pliable CRTC, no tactic should be ruled out as too presumptuous until it was first tried on the commission.

The basic part of CCL's takeover application was a 366-page, 8½-by-11-inch bound volume with a glossy dark-brown cardboard cover, which came to be known as the Brown Book. This two-pound paperback had two sections, each with five parts. Section I, Part C ("Specific benefits") in turn had ten subsections itemizing all the benefits the takeover was supposed to bring. The Brown Book came complete with light brown, typeset pages marking off the two sections, the ten parts and the ten subsections, plus, as a frontispiece, an aerial photograph, also on brown, of downtown Vancouver, Burrard Inlet and the North Shore mountains.

Section III, in a separate blue binder, weighing in at four pounds, was eight consultants' studies totalling 314 pages. Capital Cable, knowing that the lacklustre commission would never surface from underneath all this paper, wondered aloud to the commission if there were a public-hearing equivalent of an intestinal bypass.

Then came the additional legal and regulatory documentation required. The total application was an impressive 6½ inches high.

The Brown Book was full of little quotes stuck in as if in a high school valedictory, just like the application for the takeover of CCL that had worked before. It was name-dropping galore — Minister of Communications David MacDonald, President Kennedy ("New Frontiers"), Pierre Elliott Trudeau, Ontario and British Columbia ministers of communications, to name only a few. Leading off a section on strengthening the Canadian program production industry was that old Canadian philosopher, U.S. Republican George Bush ("It's tough out there," said George Bush, as quoted by CCL). Nationalist and culturally minded quotes were particularly popular.

Three of the eight consultants' studies were done by Americans. The major Canadian studies might as well have been written by Americans, too. The "market role models," as one study by a Toronto management consultant firm called them, were American (the "market role models" cited by this study were Time Incorporated, Warner Communications, Turner Broadcasting and Teleprompter). The phrase "market role models" itself was American, far removed from the political, social and cultural objectives meant for Canadian broadcasting in the face of American-dominated market forces. The innovations envisaged, such as pay-TV, were also American-inspired. Company expansion, further, was seen in American terms. Both Premier and CCL already had interests in the United States and wanted to expand them.

Several of the studies tried to rationalize the proposed concentration at great length without, however, any specific demonstration of how concentration would bring benefits and without any documentation, analysis and comparative evaluation of the disadvantages and dangers of concentration. Alternative licensee possibilities, like subscriber-owned and governed licensees, were excluded. Alternative co-operative or consortia models, with or without government sponsorship, to achieve economies of scale if and where available (technological development, for example), were largely excluded.

The application pointed out, in order to agree with, some of the obvious reasons for not allowing the takeover, which expressions of candour were meant to disarm the commission into allowing the takeover — a variation of an old confidence-game technique.

"Local ownership is usually preferable to distant ownership," it said, for example, on one of its thousand-odd pages, in one of its specially subdivided subdivisions. Obviously, any organization that was sensitive enough to point this out voluntarily, and to do so in its own application for a takeover, could be trusted with the elimination of local ownership. In the cockeyed world of CRTC doublethink, to which the Rogers application was playing, decentralization through concentration could be confidently advanced as a "specific benefit" over the more decentralized ownership already in existence!

Similarly, the Toronto application offered pandering phrases about "the sheer determination of the new West to assert itself" and other confectioned references of concern over western alienation, while it proposed, out of Toronto, to take over these western licences without westerners themselves having a chance to apply.

The astonishing success of window-dressing promises in manipulating the CRTC in the earlier takeover now spurred the Rogers officers, with the Premier officers in tow, to new and improved heights. "Research and development" provided particularly rich possibilities for dragging out lists of items and for expanding length, even if the promised increase from the takeover was marginal (0.5 per cent of gross revenue) and would be charged against subscribers anyway, through rate-increase applications. Programming matters had still more possibilities for the spinning out of pages, also for paying lengthy ceremonial respect to the "independent producers," and, best of all, for pious expressions of Canadianism — "strengthening the cultural expression of the nation," as the Brown Book put it.

The offering that made this display possible was a prominently billed "Canadian Cable Production Fund" of $5 million, put forward in conjunction with Famous Players in which CCL had a 49 per cent interest. This was a one-time expenditure only. Its carrying cost, for CCL when anybody was curious enough to work it out, came to 0.25 per cent of projected gross revenue. The Something for Something Fund was 1 per cent of gross revenue. This made a total of 1.25 per cent of gross revenue for professional Canadian program production. The CRTC was explicitly instructed that the touted production fund, plus the extension of the Something for Something Fund to Premier, would only be forthcoming if the takeover were approved. (The 0.25 per cent of gross revenue for the imposingly named production fund compared to the 10 per cent minimum, and possibly as high as 25 per cent allocation, in the Capital Cable model.)

Then there were matters that should have been taken care of by the existing licensee, like moving more quickly to rebuild the Vancouver system. "A national showcase, world-class cable system" was the application's stirring rendition of this conventional rebuild. The impression was given that only the bold and daring imagination of the Toronto corporation could manage such a revolutionary, futuristic achievement.

The fanciful proposition that cable concentration was needed to withstand future incursions by the terrible telephone companies (the wicked "telcos") — a proposition floated successfully, without drawing critical examination, for Rogers's previous takeover — was now mined for all it was worth. So was another bait used the last time: cable boundary rationalization in the Toronto area. This tactic was an excellent way to get obliging Ontario dignitaries to support the Toronto takeover of British Columbia cable licences, which they would assume would be a good thing anyway.

Rogers and his batmen then had the Brown Book reproduced in bulk (five thousand copies all told) and selectively distributed by CCL and by Premier, in southwestern B.C. and in the Toronto and Oakville areas and elsewhere, to impress important people and to scare up interventions, playing on people's naivety.

To prepare the invasion on the British Columbia side, Rogers had earlier put Peter Hyndman, then president of the Social Credit party, on the board of Rogers Telecommunications (the corporate vehicle at the time). This paid off by helping to disarm the British Columbia ("B.C. is not for sale") government. Hyndman, by 1980, was a government MLA. He later became well known as a Cabinet minister for charging a number of questionable expenses to the public purse, including a celebrated dinner at the elegant Il Giardino di Umberto, at which one of the parties was Phil Lind, vice-president of CCL, and Mrs. Lind.

Hyndman now, with the takeover imminent, approached Stuart Leggatt, a prominent NDP lawyer and also an MLA, to see if he would join Rogers's well-paid assortment of legal representatives. Leggatt had no illusions about why he had been picked out from all the lawyers in the province. Being a man of principle, he refused the offer.

Rogers (Canadian Cablesystems) then moved to capture the hearing milieu. At the back of the hearing room in the Sheraton Landmark Hotel, facing the entrance, they set up a double bank of thirty television sets, impressively stacked on each other and labelled, to represent the thirty channels of the "world-class" cable system that the takeover, and only the takeover, as was contended, could bring to Vancouver. (There was no provision in the rules of procedure for physical displays like this, least of all large wall spaces.) The

Brown Book was brought in by the carton-load and, along with glossy handouts from CCL, including a slick presentation trumpeting the takeover, was distributed from a table at the doorway inside the hearing room. It was as if the hearing was a CCL salesmen's convention. (Again, there was no provision for this distribution of literature in the rules of procedure.)

Satellite transmission of the hearing to Toronto was provided and paid for by CCL. The two-way satellite hookup was used to allow Ontario intervenors culled by CCL to make oral presentations to the hearing in favour of the takeover, whereas in the notice of hearing no such possibility was announced for those pro and con who otherwise might have intervened. The Canadian Broadcasting League, Ottawa, for example, which intervened against the takeover, was not notified of the possibility of a long-distance appearance through Toronto. The commission blissfully let this intrusion on its procedure and its possession of the hearing go by.

For the successful takeover of CCL, Rogers Telecommunications had generously catalogued and categorized all the interventions and distributed the results, trying to take over the CRTC's own work. They did it again in this case for the commission's "greater ease of reference," casually including, as interventions in support of the application, submissions that were not against the application but were not in favour of it, either. Also included were the complete replies by CCL to opposing interventions but not the interventions themselves. This handy package was then reproduced in impressive bound form under blue-coloured cardboard cover and referred to as the Blue Book. It, too, appeared by the carton-load at the special information table.

If you dismantled the elaborate and expensive package by analysis, boiled off the verbiage, tore off the coloured covers and the pasted-on nationalist piety, shook away the thick and crumbly pancake make-up of the made-to-order consultants' studies meant to cover the bad complexion of the takeover, what was left looked suspiciously similar to the two previously rejected applications by Western Broadcasting, except that Western Broadcasting had the misfortune to be from Vancouver and not from Toronto. The commitment to new Canadian program production was a trifling, throwaway come-on in both cases. The corporate orientation was American-inspired in both cases. The inner push for expansion, in both cases, was to force the introduction of American-style, American-dependent pay-TV and to get in on the expansion of cable in the United States. Oh, but the package! The straight-thinking mercenaries of Western Broadcasting did not realize that the commission was so soft-headed, and that it was worth spending the money for such a crudely artful presentation.

The first generation of private cable operators in British Columbia, led eventually by Syd Welsh and his two partners in what became Premier, were hardware capitalists. The next generation were the lawyers — president and vice-president of Premier. The Rogers (Canadian Cablesystems) crowd were the packagers. Watching this crowd, one suddenly began to grow fond of the direct, old-fashioned mercenary qualities of the Welsh generation.

CCL, in a written rebuttal, indignantly accused Capital Cable of "xenophobia towards anyone east of the Fraser Valley" for the effrontery of insisting on the right to apply for licences in their own province at the same time that a Toronto company was given the right to apply. This xenophobia, CCL's legal counsel declared, was "inimical to a united Canada where Canadians are free to move from one part of the country to another."

Apparently British Columbians were to lie down and let themselves be steamrollered, and the converse: For British Columbians to stand up for their rights was subversive. In the natural central Canadian scheme of things, which the Toronto lawyer was repeating for the commission, this made perfect sense.

Central Canadian code phrases like "east-west grid," and patriotic declarations about uniting the country, were used freely by Rogers. It was just like Al Bruner's "east-west links" in the 1973 Global sortie to the West. The central Canadian talking points in the application matched the colouration of the central Canadian agency. The hearing need never have moved west of Etobicoke. (One Ontario cable owner, Geoff Conway of CUC Limited, who had intervened in 1971 against Premier's assuming control of York Cablevision, because it was not local ownership ["Toronto is 2700 air miles away from Vancouver head office," he indignantly declared], now effortlessly intervened in support of Rogers's takeover of the Vancouver and Victoria cable licences; call it local ownership Ontario-style.) None of the consultants' studies in the application, similarly, was from British Columbia or contained anything about British Columbia, except for, in one case, a few incidental references.

The application even talked about accepting the Canadian "cultural burden" — shades of the white man's burden — that noble Canadian Cablesystems would assume in British Columbia if only the takeover were approved. The other benefits of modern civilization ("a major involvement with high technology") would also be

brought to the backward British Columbians by the Toronto mis-
sionaries. British Columbians had to be taken over for their own
good.

At the hearing Rogers and his Toronto officers presented for the
consumption of the visiting central Canadian commission a tourist-
style video clip on Vancouver as if they, Rogers, had discovered the
place and were making a gift of it to the recently arrived heads of
state which, in cable terms, they were purporting to do. The gesture
was so crude that even the commission appeared embarrassed. It
brought to mind the Wehrmacht's occupation of Paris in World War
II. It was like watching one's country being taken over, hands tied
behind one's back. Several months later, in an interview featuring
the takeover, Rogers would exultantly refer to his team as "the
stormtroopers."

The only procedural weapon available at the hearing to Capital
Cable, represented by myself and Peter Pollen, and to the APBBC,
represented by Andrew Roman of the Public Interest Advocacy
Centre, was the preliminary motion, a quixotic weapon since the
motions would be addressed to the commission itself. The commission, on these motions, ran interference for CCL just as surely as if it
were taking instructions from the sidelines.

Capital Cable and the APBBC first advanced motions that competing applications be allowed. Capital Cable underlined the commission's previous evasions on the matter.

As a prelude to its motion, Capital Cable asked the commission
who the applicant was. This sounds like an embarrassingly simple
question, but in the Orwellian newspeak world built up by the
commission around its evasion the question was subversive. The
commission was still pretending that an issuance of new licences to
CCL was not occurring here at all, only an amendment to existing
licences noting a change of control of the holding company.
According to Section 17(b) of the Broadcasting Act, however, a
licence could be amended only upon application by a licensee, in
this case, the licensee companies, Victoria Cablevision Ltd. and so
on. The commission's legal counsel was therefore obliged to claim
that these were indeed the applicants. Most of the application material, on the other hand, like the Brown Book, was filed under CCL's
name. The commission itself, in correspondence, referred to the
"application by Canadian Cablesystems Ltd."

Although Capital Cable Co-operative could mischievously point
out this contradiction and the ruse behind it, it could not guarantee
that there would be plain speaking about it. When Capital Cable

persisted, the commission's legal counsel got huffy, telling it to get on and make its motion. The poor fellow was doing his best to do what the commission's fiction required but, as a relatively new man, he was not yet in the habit of calmly stating palpable nonsense as truth. The motions were denied.

Capital Cable and the APBBC made motions for the allowance of cross-examination by intervenors of the expert witnesses — the authors of the consultants' reports. Rogers fought cross-examination hard. The motions were denied.

Roman, while the hearing proceeded, took the question of cross-examination to the Federal Court. The justice upheld the CRTC's claim to have the right to decide whether there should be cross-examination or not. This was later appealed.

The whole hearing, lock, stock and barrel, should not have taken place, not least because concentration in broadcasting and cable was properly a question for Parliament. Capital Cable made a motion to adjourn, for that and other reasons. This motion, like all the others, was denied.

When Capital Cable's turn came up on the agenda, one of the commissioners, Roy Faibish, threatened me, as the author of the organization's written submission, with a citation of contempt for a few specific phrases in the brief. The CRTC, though an administrative body and not a court, had the powers of a "superior court of record" and hence, it appeared, the powers to cite for contempt. My legal counsel, a lawyer with long experience in administrative and Federal Court proceedings, whom I consulted overnight, commented wryly that the phrases complained of were the clearest case of contempt he had ever seen. In these instances, it does not make any difference if the passages are true or not, they are still in contempt.

The next day I withdrew the phrases.

The Vancouver *Sun* ran an angry editorial entitled "The real contempt," attacking the CRTC for even trying to use its technical powers of a superior court to muzzle public opinion at a public hearing, for "puffing themselves up and pretending to be judges." The real contempt was the refusal to respond to Capital Cable's complaints, said the *Sun.*. The editorial called the commission "a pompous collection of little tin gods." The *Globe and Mail* wrote a long and stinging lead editorial, "The CRTC as censor."

Who would have guessed that just five months later, and after participating in the decision approving the takeover, Faibish would join Rogers and become a vice-president of CCL?

Back at the hearing, after the contretemps with Faibish, Capital Cable finally made its presentation. It dissected CCL's missionary pretensions and the application's speciousness, and went on from there. Rogers for the first time was thrown on the defensive; in his rebuttal, he began bringing up extraneous new arguments in an attempt to discredit the intervenor's position. This was a stock manoeuvre in such situations. No response by intervenors was allowed to applicants' rebuttals, and obliging CRTC chairmen permitted applicants to say almost anything in them. New subject matter could be introduced and manipulated with impunity.

Rogers, despite his long experience, did not have enough faith in the commission's captivity. Several weeks after the hearing Premier shares on the stock exchange mysteriously inched upward towards the $25 mark, and shortly after that the approval of the takeover was officially announced. The commission, in its decision, ducked yet again on the issue of excluding competing applications, as indeed it had to if it were to let the takeover go ahead.

In 1973 a political science professor had written in a book entitled *Working Papers on Canadian Politics* that "technological innovation and ever greater concentration of ownership and control in the economy make it increasingly difficult for private and public decision-making to occur in a manner reasonably free of domination by certain powerful vested interests." Yes, it was John Meisel. Shortly after the decision, Meisel could be seen making a profound speech weighty with references to the importance of "quality and character [in the country's] regulatory machinery"; to his concern that the CRTC function with "fairness, dispatch and openness"; to the drawbacks of takeovers and concentration and how they threatened to "homogenize" us, and to the fact that "the years ahead will require that we confront these issues."

10

The End of the Line

VARIOUS LEGAL ACTIONS against the CRTC dealing with the exclusion of competing applications in these licence takeover cases, but predating the Rogers takeover of Premier's licences, had been slowly winding their way through the courts. They were being handled by Andrew Roman of the Public Interest Advocacy Centre. One of them was the takeover of Canadian Cablesystems licences by Rogers. Another was the attempted takeover of the Western Cablevision licences by Maclean-Hunter Cable TV. The most important, however, because it was the most clear-cut, was the Courtenay-Comox case: the allocation in late 1978, just after the commission's hearing on procedure, of a new licence to a subsidiary of Agra Industries (Toronto) after the current one, held by local individuals, was surrendered. The transaction had been arranged in those terms, as a licence "surrender" and the granting of another licence. The usual pretence was that the licensee was not changing because the corporate shell remained the same, but here that pretence did not have to be contended with.

The Association for Public Broadcasting in British Columbia (APBBC) was the plaintiff in the legal action, seeking to overturn the CRTC's decision. It took a year and a half to get the case to court. The judgement, by the Federal Court of Appeal, was issued after the CRTC's Rogers-Premier takeover hearing, but before the commission's decision approving that takeover. The court action regarding this earlier, Courtenay-Comox takeover was the crucial test case for the legal defence of the right to apply in competition with rigged takeovers and the pell-mell concentration of mass media power that came with them. The case was barely noticed by the media; only a cursory dispatch or two showed up in the odd newspaper.

The plaintiff had argued that in the Courtenay-Comox situation, when a licence is surrendered and a new licence is allocated, anyone who wishes to apply should be permitted to do so, just as in a territory not previously served by a licensee. The CRTC had tried to get around this by claiming that the application was not simply for a surrender but for a surrender with a condition, namely that the licence would go to the outgoing licensee's nominee. But this, the plaintiff had contended, was in substance an application for a transfer of the licence, however it was described. The CRTC did not have the power to transfer licences. In other words, (a) if the condition was attached, the application was for a transfer of the licence and therefore beyond the CRTC's jurisdiction, and (b) if the condition was not attached, the application would be for a clean surrender, to be followed by the issuance of a new licence open to all applicants. It had to be one or the other.

In its judgement, however, the court declined to address the question of whether the takeover as arranged was indeed, in substance, the application for a transfer of the licence, and therefore beyond the CRTC's jurisdiction. It accepted the CRTC's characterization of the takeover as a surrender with a condition, *without discussing what the practical meaning of that extraordinary condition was, linked to the surrender in that way, namely a transfer of the licence.* Since, further, there was not anything in the statute specifically forbidding the CRTC from considering a surrender application with a condition attached, the commission could therefore go ahead as it had set up the application. All "natural justice" arguments (that any interested party had a basic right to apply for a new public licence) were dismissed.

This was brilliant jurisprudence, by which a transfer by any other name was taken as a nontransfer, after which competing applications could be excluded because the transaction was in reality a transfer!

There still remained one hitch. The CRTC's action in accepting a conditional surrender in that way had to be consistent with some policy aimed at implementing the Broadcasting Act. This usually would be only a technicality. For the purpose, any purported policy remotely connected to the regulation and supervision of broadcasting would do. Further, the commission had only to say that it had the policy. It was not the court's job to determine the policy's merits, realism, or even its honesty.

The sole demonstrable function of excluding competing applications, however, was to allow the dealing in licences to go on. This

private appropriation of public property without any corresponding specific authorization in the Broadcasting Act, and unrelated to
any object mentioned in the legislation, was one of the plaintiff's
major arguments against the procedure.

The policy the CRTC finally produced for the trial was that it did
not want to deprive a licensed area of cablevision service. This was
accepted unquestioningly by the court. An area might well be so
deprived of service, the court's decision explained, if the surrender
of the licence were permitted "without reference to a replacement
licence being assured."

This was a most peculiar explanation, considering that (a) if competing applications were allowed, there would be more chance of a
satisfactory replacement licensee being found, there being more
applicants; (b) in practice, the possibility of there not being a satisfactory applicant did not exist; (c) the surrender application and the
new, competing licence applications could be heard at the same
hearing and the former denied, leaving the old licence in place, if all
of the latter had to be refused, and (d) for the sake of the outgoing
licensee, it could easily be arranged for the outgoing licensee to
continue, and not be stuck with dormant assets, if by some freak
accident like a nuclear war there was no satisfactory new applicant.

The judgement also explained that an area might be deprived of
service if the surrender were first accepted "and if, thereafter,
approval to the proposed transfer [of the assets] was for any reason
refused." This was total confusion now. The ex-licensee would not
need approval from the CRTC to dispose of its assets. The new
licensee would not need approval from the CRTC to acquire the
assets, either. Further, if the CRTC were intent on continuity of
service, why would it not approve the acquisition of the assets at
hand, if its approval were required, which it was not? Why ever was
the court talking about the transfer of private assets when it was
deciding on a case about public licences?

The judgement appeared to link interchangeably the arrangement for disposing of the assets, over which the CRTC had no jurisdiction, and the arrangement for the public licence, a serious confusion by itself.

In elaborating on the policy of continuity of service, the court's
decision also included the argument that "part of the broadcasting
policy is the right of persons to receive programs." That reference
in the Broadcasting Act was a formality, presumably to prevent the
reception of signals from being unduly interfered with. In any case,
it was legally irrelevant here. A right to receive a service that is not

provided, if somehow the service were discontinued, is a right that exists only in wonderland.

The commission's proffered policy for the occasion, incidentally, also had no connection to experience. Nobody had heard of this proposition before. The commission, in its 1978 policy paper on procedure, had expressed itself *in favour* of competitive hearings in these circumstances, except that certain alleged "obstacles" stood in the way, none of which, however, was the interruption of the provision of service. How, you may ask, could the CRTC cite before the courts a policy at odds with its policy as set down elsewhere in the same period? It could do so, for legal purposes, simply by doing so. Nothing stopped the commission from having various versions of its policy and changing its mind and being inconsistent, in the free exercise of its regulatory discretion.

In licence hearings where the issue of competitive applications had been raised, and through leaks from within the commission, what was clearly evident was that the commission could not bring itself to interfere with the licensee's cashing in for whatever it could get, the value of the licence included. The outgoing licensee for its part, if it was interested in continuity of service, would have been all in favour of the greatest number of applicants. But nobody had any illusions about what the outgoing licensee's interest was in wanting competing applications excluded. For legal purposes, however, none of this apparently made any difference.

The review panel of the Supreme Court of Canada denied the APBBC leave to appeal. It gave no reasons.

Remember that phrase, direct from the Broadcasting Act, the one officials so much liked to quote, that "radio frequencies are public property"? The one that had been put into the Broadcasting Act at the order of Lester Pearson to make up for an omission in the 1958 legislation? The one inspired by the clause in the 1936 act that "no person shall be deemed to have any proprietary right in any channel heretofore or hereafter assigned, and no person shall be entitled to any compensation by reason of the cancellation of the assignment of a channel"? The court's decision in the Courtenay-Comox case, which disregarded private trafficking in the licences, wiped out any legal meaning the phrase may have had, if it had any. As one legal commentator pronounced after the fact of the decision, the declaration that "radio frequencies are public property" was merely "hortatory." In other words, it was not worth the notepaper that Pearson's aides had originally written it down on. It had been turned into a pumpkin.

With the Courtenay-Comox case down, all the other legal challenges over closed licence takeovers, under the existing legislation, also fell. Trafficking with licences originally allocated in competitive hearings continued unrestrained by any legal check. Concentration continued.

The game was not over yet, however, even under the existing legislation. There was still the commitment by then Minister of Communications Jeanne Sauvé in 1977 to take a further look at the CRTC's exclusion of competing applications. The commitment was made after Sauvé had casually rejected the Capital Cable appeal on the exclusion without having taken much of a first look.

In February 1980, a mere three years later, while privileged takeovers big and small continued unimpeded, including the takeover of Canadian Cablesystems by Rogers Telecommunications, the department finally got around to commissioning the study, during the David MacDonald interregnum as minister of communications. It appointed leading communications consultant Robert Babe and the dean of law at the University of Western Ontario, Philip Slayton, to do the job. Probably because of the imminent hearing on the takeover of Premier by Rogers, they were given a short deadline. They managed to make their submission only two months later — a comprehensive, heavily footnoted, two-hundred-page document complete with lengthy bibliography. The study, *Competitive Procedures for Broadcasting — Renewals and Transfers*, dubbed the Babe-Slayton Report by Capital Cable Co-operative, looked not only at the Canadian situation but also at licensing practices in the United States and Great Britain and at regulatory principles generally.

With regard to the expiry of licences and what would be, in a competitive process, a reallocation hearing (in the closed procedure, a "renewal" hearing), Babe and Slayton hung shy in the Ottawa winds. While putting forward competitive hearings as an option, they concentrated on other possibilities, like a surplus profits tax on the revenue generated by the public licences, as in Great Britain. These funds would then be used for independent Canadian productions to be run in peak time, without compensation, on CTV and other private stations, creating in those few hours per week a Canadian equivalent of the British Independent Broadcasting Authority. Another option was to establish a new office, entirely separate from the CRTC, which would play an adversary role in licence-renewal proceedings similar to the role played by the Combines Branch before the Restrictive Trade

Practices Commission. At the same time, the CRTC would be required by legislation to extract detailed promises of performance from each licensee and then to deny renewal if substantial compliance were not forthcoming, following which competing applications would be called for. Presumably, because of the legislated requirement, failure by the CRTC to get tough would be actionable in court.

These were no substitutes for a competitive procedure. But Babe and Slayton considered a competitive "renewal" procedure too weak by itself, since the CRTC had already let slip that it might not adhere to the spirit of such a procedure, but only to the form. The record of the Federal Communications Commission in the United States appeared to support this scenario. There, competitive hearings were required by law but the agency had virtually made licence challenges inoperative. The British example, however, was the contrary. And theoretically in Canada, as in Great Britain, broadcasting as a public service rather than as a private business was the premise. Moreover, whether or not a captive regulatory agency might undermine the principle of competitive hearings was no argument for not establishing them, only an argument for reforming the agency.

But what about the so-called transfers, in effect the allocation of new licences, which applied to the Victoria and Vancouver takeover attempt? The report called unequivocally for competitive hearings and for revisions to the Broadcasting Act to "firmly reaffirm that no private licensee has a vested interest in any radio frequency." It correctly pointed to what was causing the holdup in reform — the prices to be paid for assets — and dealt with it quickly in three short paragraphs outlining the options. It described this and its other recommendations as "very conservative."

Ottawa administration, however, resembled nothing so much as nineteenth-century Russian administration, a St. Petersburg on the Rideau, against which such good work as the Babe-Slayton Report was no match at all. The original submission of the report, requiring only minor changes to clarify a few points, predated the hearing on Rogers's Premier application by a month. It was April 1980. Capital Cable began pressing the Department of Communications for a copy to have in hand before the hearing. A new round of delays and stonewalling began. The hearing came and went without the Babe-Slayton Report seeing the light of day. Capital Cable then pressed for a copy before the CRTC decision came out. The decision, approving the Rogers takeover of Premier, was subsequently issued without the release of the report.

Capital Cable began work on its petition to Minister of Communications Francis Fox and the Cabinet, as foreseen, appealing the CRTC's decision on several grounds, the exclusion of competing applications being the leading one. The promised date for getting a copy of the report came and went without the report surfacing. Capital Cable filed its appeal. The deadline for Cabinet action on the appeal passed on September 28. On October 9, a letter went out over Francis Fox's signature denying the appeal and enclosing a copy of the Babe-Slayton Report.

The terse letter from Fox, a few short paragraphs, avoided the arguments of the appeal. The only explanation offered was that appeals on the question of competitive hearings had already gone to Cabinet twice and been rejected, magically forgetting that Sauvé had made her commitment and that the Babe-Slayton Report had been commissioned expressly because the Cabinet had not really dealt with the issue, and the report was to be a prelude to their doing so. Fox then concluded that he saw no reason for setting the decision aside and that he was enclosing a copy of the consultants' report on the issue. The consultants' report recommended, of course, competitive hearings. The report was then shelved without any release to the media or any announcement of any kind about its existence.

The person in charge of processing the Babe-Slayton Report was the acting director general, broadcasting and social policy, responsible for policy analysis on the broadcasting and cable side. Also involved in policy consideration was the associate deputy minister. Both during the summer favoured the Capital Cable appeal. At the crucial junction between them and the minister in the departmental hierarchy, however, was the deputy minister. The deputy minister, installed during the spring, after the change of government, was Pierre Juneau, the original chairman of the CRTC responsible for the very policies the appeal brought into question.

No matter what route Capital Cable took, it came to the same place in the end.

With the Rogers takeover of Premier, *Saturday Night* excelled itself. In a cover story on Rogers and Rogers Cablesystems by Toronto writer David MacFarlane, the takeover of Premier was portrayed as an unvarnished, conquering triumph. The rest of the Premier takeover story was omitted. The account floated in uncritical Toronto adulation, complete with colonizing metaphors of fur trade and railway, in which the only British Columbian quoted, a community programming employee at a fancy Rogers reception ("I know I shouldn't admit it, I don't even know who Ted Rogers is") was seen in bewildered thrall.

At the same time, in *Maclean's,* Ottawa staffer Ian Anderson did a full-length profile on CRTC chairman John Meisel. No questions were asked about the capture of the CRTC by the industry, or about the licensing scandal. Instead the reader got virtuous quotes exactly contrary to what the CRTC had been doing, and a picture of an all-powerful commission propelled by a tough nationalist, a picture that was make-believe when it was first trotted out by the same kind of journalism, with regard to Pierre Juneau, in 1970.

The Fifth Estate, As It Happens and *Sunday Morning* uttered nary a peep.

Part V

The Pay-TV Fiasco

1

The Cable Lobby Uses Its Muscle

THE PAY-TV FIASCO was not only predictable, it was predicted. Pay-TV never made Canadian sense.

As the cable industry proposed it, pay-TV would offer subscribers an optional channel, or channels, of movies, specials and other programs for an extra fee. It had the same basic drawback as conventional commercial television: It depended on buying American. The Canadian pay-TV market was too small to finance enough high-quality Canadian production to attract a large enough audience to keep the scheme going. Only American films, acquired at a small fraction of the production price, and carrying the marketability of American publicity and American stars, could keep the operation viable. The economics of regional pay-TV and of French-language pay-TV were even worse than for national English-language pay-TV. Financially speaking, Canadian pay-TV was a contradiction in terms. You could hold public hearings on it for a lifetime, around the clock and on weekends, accepting and discussing submissions and boosting broadcasting lawyers' incomes, and what would come out at the end would be (a) another extension of the American distribution system and (b) an increase in our balance-of-payments deficit with the United States.

These were not characteristics that would help the cable lobby get pay-TV through Ottawa. So it began to talk about the great benefits that pay-TV would bring to the Canadian program production industry, particularly to the independent producers. The lobby in 1975 offered 15 per cent of gross revenue to Canadian production. By using a five-year instead of a one-year figure and also the most optimistic projection, the lobby then magically inflated the projected production expenditure from $4 million to $80 million, which sounded better. Another 25 per cent was indicated for American and other foreign acquisitions. As in conventional television, these foreign acquisitions would be made at a small fraction of their production cost. The cable companies who, in the scheme, would be delivering the channel, required a minimum 55 per cent of gross revenue, largely, they claimed, for the special hardware to provide the signal only to pay-TV subscribers. The remaining 5 per cent would go to miscellaneous costs, including pay-TV profit margins.

Like conventional commercial television, these overhead costs were extraordinarily wasteful. In this case, 85 cents on the dollar would be spent on the apparatus and on imports, and only 15 cents on the dollar for the Canadian program production itself which, according to the pious declarations, was the major justification for pay-TV. The way to finance new Canadian production was to finance it directly — just do it. In public funding mechanisms or in the Capital Cable Co-operative model, moreover, there were ways of doing it with virtually no financial overhead costs. In most businesses, doing something at five or six times, or even two or three times, the necessary cost would be considered absolutely screwball. In the captive Canadian broadcasting policy business, in Ottawa, it was taken as the height of high finance, which it was, for the cable companies.

There was another built-in perversity in pay-TV. It was not universal. The aim of all the CRTC hearings, important-sounding ministerial statements, multifarious demonstrations of hair-pulling, the media coverage and the rest of the vast elaborate exercise surrounding Canadian broadcasting policy was to increase the presence of Canadian programming in the face of the American programming bombardment. If money was going to be spent on Canadian programming out of the economy, then why not distribute the programs to everybody off the top, the way it should be done, instead of only to a minority, or only after a delay, on a second-rate basis?

The most revolutionary aspect of radio and television broadcasting was its universality. Television had gone a long way towards creating an "open movie house." A pay-TV channel would be the first step in breaking this universal access down for future television and information services, to the greater profitability of cable companies and marketing bureaucracies.

One of the original arguments for pay-TV advanced by the cable lobby was the absence of commercials. The lobby even tried to make out that one of the functions of pay-TV was to recover Canadian money going down to the U.S. in contributions to noncommercial PBS stations from places like Vancouver and Victoria. The absence of commercial interruptions was not an argument for pay-TV, however. It was an argument for public noncommercial television. Pay-TV would create a situation where a small minority, with extra disposable income, would have the luxury of television entertainment uninterrupted by commercials, whereas Canadians as a whole, who were already paying through a public mechanism for most of the CBC's budget, would have to continue to endure commercials on their publicly owned network. They would have to continue to endure commercials on private stations which, indirectly, they all financed too through consumer purchases.

Another argument that came up in the original days of the pay-TV lobby was that pay-TV would introduce specialized programming not currently available on conventional television. Opera, ballet, drama and symphonies were particularly worth mentioning. People talked about getting the Royal Winnipeg Ballet, the Toronto Symphony Orchestra and the Stratford Festival on pay-TV. The economics of pay-TV, particularly in Canada, however, were just like those of conventional commercial television. They pushed remorselessly in the opposite direction: mass-appeal programming. Specialized Canadian programming would not have a chance, least of all expensive productions like opera. It was obvious from the beginning that a pay-TV channel would basically be an American movie channel, with some specials and sports programming thrown in. Later, pornography would come on the scene.

Even if pay-TV by some unforeseen development were able to manage cultural and similar specialized programming, it would be the wrong place for it. By far the best and the cheapest way of broadening public taste with specialized programming, of developing new viewer interests and of giving our artists and innovative producers the largest possible audience, is to make the programming universally available. Public broadcasting — the CBC, PBS, BBC and other public broadcasters — do provide this programming variety. The way to increase the amount of it is to increase the financing and the number of channels of public broadcasting.

Except for producing a bit of employment, moreover, the projected 15 per cent of gross revenue for Canadian production did not mean much. The economics of this production would oblige it to find American sales, and to Americanize its production accordingly, just like the notorious coproductions and other similar arrangements for U.S. sales undertaken by conventional commercial television production in Canada. The need for these "international" sales for Canadian pay-TV production was understood and usually explicitly cited in discussions about pay-TV.

Canadians in any case were not demanding pay-TV. One early survey in 1976, where the concept was explained, found that only 8 per cent were very interested in pay-TV. Another 16 per cent were quite interested. A large majority, 72 per cent, were either not too interested (25 per cent) or not at all interested (47 per cent). A CRTC survey done in late 1977 showed that most Canadians knew little or nothing at all about pay-TV and that, when it was explained to them, only 3 per cent said it was very likely they would subscribe to the service on a channel subscription basis at $8 per month.

Trying to re-Canadianize television with pay-TV was trying to re-Canadianize it by Americanizing it even more thoroughly. This absurdity was no secret. The cable industry's pay-television lobby was frank about it. "We see two kinds of people making money with pay-TV," Colin Watson, the head of the lobby, said. "One, the cable operators . . . Second, the Canadian production community — and the cost of that is inundating pay-TV with American programming. If there isn't recognition of that, there isn't going to be pay-TV."

Canadianization through Americanization? The minister of communications's and the CRTC's capacity for evasion was up to it. Picture the situation. The minister and the CRTC will not meet pressure from the pay-TV lobby head-on with a vigorous public broadcasting strategy. They have to avoid, then, facing up to the dominant factors in pay-TV. Otherwise there is no way out. They resort instead to bureaucratic manipulation, including delay. This manipulation cannot in the long run do much good, but it does allow for the necessary covering talk accompanying evasion, just as the rhetoric of the Fowler reports in the 1950s and 1960s helped those two commissions to indulge in basically the same kind of evasion concerning conventional private television.

The nationalist eloquence was memorable. Nationalist eloquence was the key to the "Canadianization-through-Americanization" pay-TV policy.

"If pay television is to be introduced into Canada it must result in greatly increased opportunities for Canadian creative talent and significant development of the Canadian broadcasting and program production industries," said the CRTC, for example, in 1975, on the occasion of the first commission hearing on the subject. It found the introduction of pay-TV in Canada to be premature, but opened consultation on how it could be done "consistent with" Canadian objectives.

Shortly after, Minister of Communications Jeanne Sauvé, in a speech to the Canadian Cable Television Association (CCTA), announced that pay-TV was "inevitable." Sauvé had been spooked by a small closed-circuit movie service starting up in a Toronto apartment and townhouse complex, and by threats from the cable industry, led by Ted Rogers, to follow suit. She called for pay-TV proposals. She declared the prospect of pay-TV to be the most exciting of new services, a "watershed in the development of [Canadian] broadcasting."

CRTC chairman Harry Boyle also now declared that pay-TV was inevitable because of technology and pay-TV's growth in the United States. Boyle also held private talks with the two large Toronto cable companies who were threatening closed-circuit action, in keeping with the CRTC's practice of consultation. He tried to turn the policy changeabout to Canadian advantage. Speaking to the CCTA after Sauvé, he described her pay-TV announcement as the beginning of a national policy of "cultural security" for Canada and spoke of the introduction of pay-TV as a serious first step towards the "repossession of a Canadian broadcasting system."

This Canadianizing fantasia duly rubbed off on commentators, not least nationalist ones, in inverse proportion to their knowledge. After all, went the deferential assumption, if Sauvé and Boyle were making great Canadian statements, there must be something great and Canadian in the prospect.

"Pay-TV can do a lot for Canada," announced an editorial headline in the nationalist Toronto *Star*. "It is an opportunity Canadian talent has been waiting for."

"Once tamed, pay-TV may be the best friend Canadian culture ever had," said a headline in *Maclean's* over an article by Martin Knelman. "Sauvé's intention," wrote Knelman, "is to head off further dumping of American mass culture in Canada and at the same time to strengthen Canadian production." And under a picture of Sauvé: "De-Americanizing the painless way." This line was just right for describing an extension of the American distribution system, in *Maclean's* inimitable style of profound analysis.

"Doctor, I have cablevision," wrote enthusiastic nationalist Jim Bacque about pay-TV in *Weekend Magazine*.

Regulars were not so gulled by the prattle. The most passionate in hitting at the Americanizing implications of pay-TV was Blaik Kirby of the *Globe and Mail*. The most cynical was Don Stanley of the Vancouver *Sun*. Stanley, who was a practitioner of *journalisme vérité*, had a fine eye for the details of hypocrisy, put-on, greed and manipulation. Every time a cable lobbyist pushing pay-TV came out of the Toronto or Ottawa woodwork, Stanley would dissect the lobbying with chronicler's fervour. The pay-TV story, he knew, was a story about lobbying.

Jean-Claude Leclerc, editorial writer for *Le Devoir,* picked up Ottawa's escapism in a trenchant article entitled "The mirages of 'pay' television." "Instead of raising the requirements of the many networks that they have extended at great expense across the breadth of Canada," he wrote, "the Canadian authorities in communications mask their failures behind shiny new paradises. . . ." Leclerc correctly dismissed the official "debate" on pay-TV as a phony one. There was no use waiting for reform from Ottawa, he added. "More than such sinkholes [as pay-TV] are going to be necessary to create the basis of a national policy of 'cultural security' piously set forth by Madame Sauvé."

These churlish newspaper columnists could write all they wanted. The CCTA by 1976 had eleven people in its office, plus the executives of its various member companies, for organizing and lobbying purposes, all funded involuntarily by cable subscribers through subscriber revenue collected under public licence. The leading CCTA operative and public spokesman was its president, Michael Hind-Smith, an ex-vice-president of programming at CTV. The CCTA was an American-style lobby — slick, a strong mouthpiece for the major companies — that made the Canadian Association of Broadcasters look old-fashioned by comparison. Members of the CCTA, following Sauvé's announcement, formed the Pay Television Network (PTN) to put together a proposal and head the pay-TV lobbying. PTN launched a public relations campaign, including a glowing pamphlet for cable subscribers. The pamphlet urged subscribers to write to Ottawa.

A coalition of production, performing arts and public-interest groups that eventually totalled 220 organizations demanded that another public hearing be held on the general issue before the CRTC went any further. This and other pressure did force a hearing, the second one on pay-TV, in 1977. A total of 140 written submissions were sent in, a majority of which were against the introduction of pay-TV.

The CRTC, in its report on the hearing, found that no single proposal met Canadian broadcasting objectives. It postponed the introduction of pay-TV again. It called for "a predominantly Canadian programming service of high quality" and put forward guidelines: 50 per cent Canadian content, and the expenditure of 35 per cent of gross revenue on Canadian production. Since "a predominantly Canadian programming service of high quality" made no financial sense for pay-TV, the CRTC's report was interpreted by many as an indefinite delay of its introduction. As PTN knew, however, the report represented the normal progression of lobbying and response. By citing specific figures, the report had accepted the lobby's categories. All that was required now was to show that the requirements were impractical, upon which they would be scaled down.

The main weapon of the lobby was steady propaganda about pay-TV, maximum media coverage and letters to the editor, and co-opting as much of the hungry film- and program-production industry as possible. If Americans could have pay-TV, with newer movies for which they would otherwise have to wait a year or two to see on conventional television, why could not Canadians? The lobby got good leverage from talk about the pressure of the American pay-TV phenomenon, pressure which it was itself largely responsible for generating, helped by hefty lobbying funds. And because their pay-TV proposal would at least put something into Canadian production — unlike the Saskatchewan closed-circuit operation and other ominous possibilities, it was pointed out — leading pay-TV lobbyists could paint themselves in patriotic colours as ardent nationalists helping to save the country's soul.

The spectre of Canadians buying satellite dishes to get American pay-TV, without a care for Canadian content, now appeared on the Ottawa scene. Unlicensed satellite dish operations were sprouting up in northern communities, pulling in U.S. satellite pay-TV and other U.S. satellite channels, and cocking a snook at Ottawa and its regulations. The Department of Communications was discombobulated by this anarchistic, disrespectful development. Other satellite dishes also materialized here and there. In truth, the number of people receiving U.S. pay-TV this way was statistically insignificant, and public demand for it remained slight; isolated communities had resorted to satellite dishes simply because their choice of channels had been so limited. And the threat of U.S. pay-TV covering the country via cheap and small home satellite receivers was self-cancelling. When the dishes became cheap enough to proliferate in Canada, they would also proliferate in the United States, and the U.S. pay-TV networks would scramble their signals. But the spectre remained in the confused Ottawa air anyway.

At this stage, David MacDonald replaced Jeanne Sauvé as minister of communications. MacDonald was known for his strong social

conscience and for his impassioned support of cultural development. He was a man who had the courage of his convictions, and
was a cultural nationalist. He was also jocosely referred to in industry circles as a dolphin among a school of sharks. "The betting is on
the sharks," said the *Financial Post*.

MacDonald proposed the introduction of national pay-TV and
pushed its advent one irreversible step further by arranging for
another round of CRTC hearings on the subject, this time leading up
to a licence hearing. MacDonald's letter to the CRTC on pay-TV
elaborately spelled out seven objectives and enunciated nine guidelines. "Pay-television offers a new and unique opportunity to support the Canadian broadcasting system," he informed the CRTC. It
was a unique opportunity, all right.

The third round of hearings on pay-TV was conducted in 1980
by what came to be known as the Therrien Committee, after CRTC
vice-chairman Réal Therrien. Its primary mandate was to make
recommendations for the extension of service to northern and
remote communities. The pay-TV question was awkwardly
attached. The members of the committee included non-CRTC provincial and northern representatives. The lobbies showed up again,
for another run-through. The downhill slide continued.

The Therrien Committee filed its report. In a split opinion, with
a minority of committee members dissenting, the report concluded
that pay-TV would offer "a new and unique opportunity to foster
the beneficial development of the Canadian film and program production industries." There was the old "new and unique" gambit
again. The report elaborately made fourteen pay-TV recommendations. The minority opinion, "that no significant benefits would
accrue to the broadcasting system, and that there is little real consumer demand for pay-TV," and that pay-TV would be basically
another "pipeline for still more U.S. programming," was overruled
with a shiny Pollyanna description of the boon to come.

Despite the Therrien Committee's pay-TV go-ahead, the cable
industry, led by Ted Rogers, kept up the pressure, becoming more
truculent. It argued that the cable industry would lose apartments
to satellite dishes (and American pay-TV) in a big way unless it had
pay-TV of its own quickly. It wanted a temporary licence immediately, and it made what the *Financial Post* deftly described as "not-
so-veiled threats" to start closed-circuit pay-TV, or for some apartments to install dishes and retail U.S. pay-TV through subsidiaries
— out of necessity, it said, to compete — if Ottawa did not act
forthwith.

Now Francis Fox was minister of communications and John Meisel was chairman of the CRTC. The rubber-stamping of the Rogers takeover of Premier, the crude suppression of a possible Capital Cable Co-operative licensee and the unannounced shelving of the Babe-Slayton Report had just taken place. It was late 1980.

The CRTC commissioned a sequel to the 1977 survey which had indicated that pay-TV was not considered essential and that there was no compelling demand for it, even among cable subscribers. The new survey showed that despite the growth of pay-TV in the United States and despite, particularly, the campaign of the Pay Television Network and the rest of the cable lobby, and all the publicity they, their allies, journalistic curiosity, ministers' statements and the CRTC hearings had generated, only 14 per cent of cable television viewers said they would be "very likely" to subscribe to pay-TV at $9 a month (the price would turn out to be $16 a month), while 50 per cent said it was "very unlikely" they would be interested. Another 14 per cent said they were "somewhat unlikely" to subscribe; 17 per cent were "somewhat likely." Of total television viewers, including noncable subscribers, only 10 per cent said it was "very likely" they would subscribe to pay-TV at $9 a month on top of the regular charge, while 56 per cent said it was "very unlikely."

The CRTC called for applications anyway, in April 1981. "The Commission . . . invites applicants to be as imaginative and innovative as possible," the announcement elaborated. "Applicants are urged to respond to the particular social, cultural, linguistic and geographic characteristics of Canada and to propose original ideas which draw on the Canadian experience. Applications particularly suited to the Canadian context will be preferred to those merely imitating existing modes of pay television." The commission, similarly, was determined that pay-TV provide, yes, a "new and unique" contribution to the Canadian broadcasting system.

The call asked for regional and local applications as well as national ones. It envisioned, it seemed, a liberalizing pay-TV revolution whereby "currently neglected or under-utilized sources of Canadian programming . . . unable to gain access to the broadcasting system" would rise up in grand display. "Strengthen the Canadian broadcasting system" and "increase the diversity of programming available to Canadians" were also featured phrases. This maximum use of ritualistic language by the CRTC helped to generate and reinforce the self-deception necessary — helped to reach the necessary level of fantasy — in order to do something that the cable lobby had pushed on Ottawa but that could not be carried through to a licence decision in all its nakedness. It followed the Third Law of Regulatory Agencies: The more elaborate the ritualistic language used by the agency, the more dubious are its actions.

"The opportunities provided by pay-television must enhance our cultural life and boost our program-production industry to a position of world-class excellence," said CRTC chairman John Meisel weightily, conforming to this Third Law.

The licence hearing took place during a marathon three weeks in September and October 1981.

"Through this, his biggest decision, John Meisel sat almost Solomonic above the fray," wrote Ian Anderson dramatically in *Maclean's*. "The commission was in disarray when he took over, six of the nine commissioners having left before their terms were over. It was a rather timid former academic who then spoke of the 'mind-blowingly important issues' he faced as chairman. Now his grasp of the industry is disputed by nobody ... Behind the bemused smile ... fidgets the same mind that once concluded that Canada's 'progressively and ultimately annihilating Americanization is in part a consequence of our milquetoastian nationality.' If ever John Meisel had the chance to do anything about that 'Americanization,' the pay-TV decision is it." Presumably, all he had to do was to choose the right applicants.

You couldn't beat that for fantasy, in the *Maclean's* tradition. "... A television revolution," noted the headline over the article. Anderson was the one who had done the profile of Meisel in *Maclean's* a year earlier.

On the other hand, Robert Fulford, in a lengthy analysis in *Saturday Night*, wrote a blistering attack on the "rhetorical overkill" and the rest of the sham of the hearing. "What this [CRTC and applicants'] grandiloquence means in practice is that an aura of chicanery and dissimulation once again surrounds Canadian broadcasting," he commented.

The scene of the chicanery and dissimulation resembled a rush for gold, with some of the old performers making another burlesque turn. There were twenty-seven applications heard in all. Foremost among the applicants was Moses Znaimer of City-TV, whose consortium (Premiere) included lawyer and Liberal operative Jerry Grafstein (another founding shareholder of City-TV), former Premier cable executive Gordon Keeble, former CRTC vice-chairman Jean Fortier, and media and publishing names Jack McClelland, Anna and Julian Porter, Peter Gzowski, Alexander Ross, Peter Newman, Norman Jewison and Patrick Watson, among others. Al

Waxman, Louis del Grande, Monty Hall and no less a national personage than Alan Eagleson himself provided endorsements. Znaimer, said an admiring report, "says things about Canadian patriotism that lead his teammates to talk about him more as disciples than fans." His syndicate's presentation at the hearing opened with a nineteen-minute film that included clips of people calling him brilliant. Pay-TV was a new dawn. Znaimer was known to say things about business theory that moved people to profound admiration, one report said.

CTV, featuring veteran CRTC-massager Murray Chercover, had an application in combination with TVA, its French-language counterpart. So did Standard Broadcasting, owner of the CTV station in Ottawa and many other broadcasting properties. Standard was controlled by Hollinger-Argus, headed by Conrad Black. Philippe de Gaspé Beaubien, head of Télémédia Communications (radio and television properties) and owner of *TV Guide,* joined with Harold Greenberg of Astral Bellevue Pathé (a film distributor) for an application. Greenberg was also a film producer who had mastered the art of pasting together productions with non-Canadian leads and non-Canadian stories that nevertheless qualified for Canadian tax-shelter purposes (in other words, were heavily subsidized by tax expenditures). Beaubien "gets so worked up with nationalistic fervor, telling about his dreams for Canadian programs," read one newspaper description, "that he's been known to leave a room, suddenly, in mid-phrase, to compose himself."

Edgar Cowan, the president of Lively Arts Market Builders, another applicant, was a founding shareholder of City-TV. Don MacPherson, president of an application group called First Choice Canadian, was a former Global, CBC and CTV senior executive. His right-hand man, steering the application through the shoals, was Peter Grant, a communications lawyer who in earlier years had acted as special counsel to the CRTC and had sat on the commission side in some hearings. Hugh Faulkner, former secretary of state and now a vice-president of Alcan, had been recruited by Performance, another syndicate with an application. The leader of the Performance syndicate was Jack McAndrew, former head of CBC-TV English-language variety, one of whose great "Canadian" inspirations was a Wolfman Jack series.

These were just a few of the cast of many hundreds. They were backed directly and indirectly by well-known financial and corporate operators in addition to the Blacks, like the Eatons, the Bassetts, Harrison McCain, Maurice Strong, Edward and Peter

Bronfman, Laurent Beaudoin (Bombardier) and others quite rich but less well known, as well as by pension funds, life insurance companies, a fish processing company, a handful of oil men and an oil-rich Indian band. "We're in the world series now," Canadian Press reported one gawking observer as exclaiming.

Not too many people appeared to pay attention to Saskatoon television executive Ian McCallum, the head of Cablecom Corporation, the small closed-circuit pay-TV service already in operation in Saskatchewan. He told the CRTC hearing that many of the major applicants were "walking bankrupts and don't know it." He warned that people by and large would only pay for a television channel if it were almost all new movies and a few specials. He also said that pay-TV had quite a few adult movies which could shock people if they were not educated in advance. As for Canadian production, Cablecom, despite its genuinely wanting more Canadian material, financially had no leeway, least of all for regional programming where the market was much smaller than even the small Canadian market. "We haven't been able to initiate any production," said McCallum. "We have had no impact whatsoever on the development of Canadian production." "Regional" pay-TV, it was clear, was just imported pay-TV regionally selected. Cablecom had been programming 4 per cent Canadian content, with ambitions to raise it to 7 per cent.

In March 1982 the CRTC awarded a national general-interest pay-TV licence (French and English services) to First Choice Canadian; a national specialty (performing arts) licence to Lively Arts Market Builders; regional general-interest licences for Alberta, Ontario and the Atlantic region, and a regional multilingual licence for British Columbia. The details were spelled out in a majestic ninety-four-page decision, with elaborate appendices laying down conditions for each of the licensees.

The major figure in the regional pay-TV licensing was Dr. Charles Allard, the Edmonton ex-surgeon who already owned CITV in Edmonton. He was given the Alberta licence. He also had a 46 per cent interest in the successful application for the Ontario licence. These would both come on stream as Superchannel. Allard was the one who, explaining the need to use American stars in his "Canadian" program production, had compared the economics of television production to chemical plants. He also confessed that he had got the CITV licence without knowing the business, in explaining why, almost from the start, he had been obliged to abandon the programming for which CITV was licensed. Allard was duly punished by the rigorous CRTC ("I can assure you we are going to move very carefully and make sure people do what they say they are going to do," intoned John Meisel) by being given a big chunk of pay-TV territory. This pay-TV territory, too, was for regional purposes — to allow for the expression of distinctive regional imaginations — not exactly what a chemical-plant man like Allard had in mind. On the other hand, for Canadianization through Americanization, Allard was perfect.

Canadian content is "the whole purpose" of pay-TV, said John Meisel, summing it all up. "It's our last chance to get Canadian content right." And: "What we are doing is to help the genie out of the bottle so that Canadian productive genius and skill will be fully employed to delight Canadians."

"Handing over the chalice of Canadian culture for protection," echoed *Maclean's*, in its eternal journalistic fantasyland, under a cut of Meisel at the microphone announcing the pay-TV decision. The headline described the pay-TV licensees as "the new purveyors of national culture."

2

The Bubble Bursts

ONCE THE CONJURATION of lobbying and the necessary reciprocal self-deception of CRTC licensing was over, pay-TV in Canada could proceed as U.S.-pay-TV-in-Canada, as it was bound to do, and which it did like clockwork.

The price per channel came in at $16 per month, not the $8 and the $9 used as a measuring stick in the 1978 and 1980 surveys which even then had shown little demand for pay-TV.

First Choice and major regional channels (Superchannel in Ontario and Alberta) floated in on an ocean of glossy and expensive hype, flogging American movies and stars (some of the same movies were carried by both syndicates; they also both opened on February 1, 1983, with *Star Wars*). In its opening push, U.S.-pay-TV-in-Canada, including participating cable companies, spent, according to an unnamed cable executive cited in the *Globe and Mail*, $18 million on propaganda, not stuffing U.S.-style pay-TV down our throats — pay-TV, after all, was touted righteously as a "discretionary" service — but brainwashing us so we would then stuff U.S.-style pay-TV down our own throats. A similar kind of propaganda expenditure was envisaged for the fall. (It was later revealed, after eighteen months of operation, that 45 per cent of all revenue had been spent on marketing.)

C Channel, the "cultural" channel offered by Lively Arts Market Builders — which, however, also used movies (they tonily called it cinema, as in "Cinema," "More Cinema" and "Even More Cinema") — quickly went bankrupt in June 1983, only four and a half months after launch date. Not many people were aware that the Lively Arts syndicate had lobbied with intensity against the licensing of a CBC-2. A CBC-2 would have brought more of the arts to the general population, for which artistic activity, through the Canada Council and other grant systems, the public was already paying a share. It would have dispensed with the wasteful overheads of pay-TV. It would also have provided a French-language service (C Channel was available in English only). But then, a CBC-2, by occupying some cultural programming territory with a separate channel, would have also laid low the plans of the Lively Arts syndicate in Toronto to get in on pay-TV.

Bankruptcy could not have happened to a more deserving bunch of guys.

Regional pay-TV that was regional was a dead duck, too, by the law of economic incentives. The major "regional" licences, Superchannel in Alberta and Ontario, were U.S. movie channels like First Choice except more so, offering more movies than First Choice itself. The natural economic interest of these "regional" licensees was to pool their programming — in effect creating another national licensee — for bidding for American movies and for spreading out production costs. The Alberta and Ontario licences were tied together through Charles Allard to begin with. Later, in July 1983, Allard was also given, for lack of other contenders, the pay-TV "regional" licences for Saskatchewan, Manitoba and the Northwest Territories.

In February of that year a British Columbia–Yukon licence was granted to Aim Satellite Broadcasting Corporation. It began making informal co-operative arrangements with Superchannel. One of its major investors, however, seeing the financial writing on the wall, refused to come across, so the British Columbia–Yukon licensee never got off the ground. Allard, with CRTC approval, and to the great relief of the original promoters, took over this licence, too, and began sending his Alberta signal into British Columbia.

Half a year into operation, in August 1983, Finlay MacDonald, the president of Star Channel, the Atlantic region's licensee, told the press that Star Channel "has held its own in the marketplace and is not in trouble despite rumors to the contrary." Star Channel went into receivership a few months later.

A "regional" licence for Quebec and adjacent provinces was also awarded, to a syndicate called TVEC, to add to the French-language national service of First Choice (Premier Choix). There had been no applications for the regional licence in Quebec the first time around. The idea was to give French-language pay-TV in Quebec parity with English-language pay-TV elsewhere. Six months into operation, both Premier Choix and TVEC were on the ropes, running up large losses. "It's a phenomenal catastrophe," was how a

Quebec author of a book on pay-TV put it. The TVEC people made an appeal to the Quebec government for help. In the end Premier Choix and TVEC were combined and refloated with the help of $3 million from a Quebec crown agency and another $3 million in loans coguaranteed by the agency. For all the fuss, French-language pay-TV in Quebec was U.S. pay-TV just like its English-language counterparts, except for a greater use of films from France and other European countries. The nationalist (in the provincial sense) Parti Québécois government ended up financially supporting U.S. cultural inroads just the way that the nationalist NDP government in Saskatchewan had ended up doing the same with its rescue pay-TV operation.

At the same time, First Choice slid to the edge of receivership, to be rescued at the last moment by Astral Bellevue (Harold Greenberg) and Edward and Peter Bronfman, with the help of an emergency CRTC hearing and an obliging overnight decision resembling the Global rescue in 1974.

"Our investors have deep pockets," First Choice president Don MacPherson had said importantly when the licence was awarded. But they were not deep enough. MacPherson, whose personal credentials had been responsible for First Choice originally getting the licence, shortly after took a walk.

The CRTC, in addition to regulations for the hours of Canadian content, had for the first time, in its pay-TV decision, laid down conditions for a minimum expenditure of total revenues on Canadian production. This looked impressive. First Choice, for example, had to spend 45 per cent of its total revenue on Canadian programs compared to the meagre, nominal 15 per cent offered in 1975 by the cable lobby and considered so appalling. But unlike the cable industry figure, the CRTC requirement was calculated on a base that did not include the cable company's portion of subscriber revenue. Including that portion, First Choice's Canadian production requirement amounted to only 25 per cent of the subscribers' outlay. Only half of that, however, or 13 per cent, needed to go to "dramatic" productions and the so-called independent producers that the pay-TV superstructure had supposedly been built to serve. The rest could be spent on programs like hockey (a Superchannel feature; hockey also helped conventional stations like BCTV, CITV and CKND to fill their Canadian programming requirements). All that this minimum requirement meant, in turn, was that it had to be spent on programming made in Canada — which in practice meant that programming would likely be tailored for U.S. sales.

Superchannel in Alberta similarly was required to spend 35 per cent of its revenue on Canadian programming in total, which would work out to approximately 20 per cent, or one-fifth, of what was paid out by subscribers. For Superchannel in Ontario, the corresponding expenditure for Canadian programming would come to just over one-quarter of subscriber payments for the service, and for C Channel, the minimum required expenditure would come to approximately one-ninth of overall subscriber payment.

Shortly after the licensing decision in March 1982, production of Canadian content that could sell directly to U.S. pay-TV, as well as being offered to pay-TV in Canada, got underway. The very first "Canadian" pay-TV drama series to go before the cameras was in this vein. The series was called *33 Brompton Place,* variously described as "sex-laden," "an adult *Dallas,*" "a nudie" and "a weekly sex-bubbling soap opera," one of whose features was that you not only saw a couple in bed making passionate love, you also saw them first take off their clothes and jump into bed together. Or so preview reports had it. "It changes the chemistry," said the producer-director proudly.

This "Canadian" program had all the classic features. It was set in Chicago. Its original production depended on a presale to the United States (in this case, five episodes to the pay-TV network Showtime), and it would rise or fall according to future U.S. sales. The series was concocted by an American production company, which then arranged a coproduction with Global, which planned to run the series after its pay-TV use. The flesh and the studio location were Canadian. The lead character was Bill, a Playtime editor who was an alcoholic and into kiddie porn. "You'll learn to love him," the producer-director joked to the actresses at the audition.

Although *33 Brompton Place* did not manage to sell to pay-TV in Canada, the First Choice pay-TV network came up with a deal of its own in the genre. Instead of a mere make-believe Playtime editor, First Choice announced in January 1983 that it would offer real Playboy movies and other Playboy Channel material, touching off a controversy about pay-TV pornography. In the middle of it was Phyllis Switzer, vice-president of programming. Switzer was one of the two original principals of City-TV, which in its early years had used baby blue movies to boost its audience and revenues. She had been brought in by First Choice to arrange coproductions with American pay-TV and cable networks. Now she had to answer questions even from her sixteen-year-old daughter. ("She asks me, do I really believe in what I do — I always have to say yes," said Switzer.)

By arranging for production of Playboy material in Canada with Canadian casts, First Choice got itself some extra Canadian content in the "drama" category. Some of the work would be done through the facilities of Toronto's Glen-Warren Productions, the production arm of CFTO (Baton Broadcasting), CTV's flagship station. Other production was to be undertaken by Télé-Métropole, owner of the Montreal commercial French-language station and the power behind the TVA network. A Télé-Métropole executive admitted that the movies would be made "to their [Playboy] specifications." First Choice contributed hefty licence fees to the Playboy Channel for shows that met Canadian-content requirements.

Premier Choix and TVEC also resorted to what was variously described as "erotica" or "pornography" — including a multiple run by Premier Choix of a hard-core New York Erotic Film Festival featuring, among others, a lady using a cucumber — shows which generated diverse criticisms and protests.

The Playboy controversy was just a small part of pay-TV's Americanization of Canadian content. A leading example of "Canadian" production given by Superchannel, in the summer of 1983, pay-TV's first year, was the movie *Louisiana*. One of First Choice's leading examples of "Canadian" content the same month, August, was a two-part series, *American Caesar*, a documentary of the life of U.S. Gen. Douglas MacArthur.

"The national availability of First Choice Canadian in both languages across the country will put [in place] an organization of size and scale to match and outrun its rivals to the south. What a nation *can* determine and originate, it *can* control." That was First Choice's pledge of support for Canadian production. Now that they had the licence, the shameless enticing of American participation was the game. "We're giving out money — would you like some of it?" Riff Marcowitz, First Choice senior vice-president told a writer for the *Hollywood Reporter* in late 1982. "It's not, let's go to Canada and save money. It's, let's go to Canada and make much more money."

According to a study done by Toronto lawyer Douglas Barrett for the Canadian Conference of the Arts, released in the spring of 1983, financial and creative control of most new "Canadian" productions generated by pay-TV was held by non-Canadians.

The finest manoeuvre of all, however, was scaffolding which came up in connection with the Playboy and some other First Choice deals. It worked like this: Normally a U.S. producer would approach a company in Canada, or vice versa, suggesting a coproduction with a minority, usually 25 per cent, Canadian interest. But in scaffolding the arrangement would be written up instead as a 100 per cent Canadian-financed production with a subsequent 75 per cent export sale to the American producer. The net Canadian outlay was the same in both cases, 25 per cent, but with scaffolding, the entire investment for the production could be entered on paper as a Canadian programming expenditure. Of course, the program content would be essentially American. Pay-TV companies in Canada vehemently denied they would use scaffolding to meet their Canadian-content expenditure requirements. But legally, when it came to their conditions of licence, there was nothing to stop them, sending the CRTC back to the drawing boards.

This pay-TV "revolution," primed, fantasized, hyped, injected, flogged, peddled — it received an extraordinary amount of publicity in the process — was really what it always was, a straightforward, prosaic, altogether predictable, mercantile, high-cost extension of the American distribution system in Canada. The licensee companies and the cable lobby acted with the same eager collaboration as the original compradors in nineteenth-century China.

"We reserve the right to set matters right, through the toughest regulatory action available, in the event that this becomes necessary; we mean business!" said John Meisel toughly when announcing the pay-TV licences. U.S.-pay-TV-in-Canada then proceeded to give Meisel and the rest of his hopeless CRTC cast the business.

3

Universal Pay Television

THE ONLY WAY

WHILE THE CRTC committed its pay-TV botchery, there was, all the while, a workable alternative. The CRTC avoided implementing it.

This alternative was a subscriber cable channel available to everybody on cable. It involved the allocation to a new, largely Canadian entertainment channel of up to $2.50 per month from the basic cable charge of all subscribers. The concept came to be known as "universal pay television." Lo and behold, it was the "colicensee proposal" first put forward by the Association for Public Broadcasting in British Columbia in 1973 and 1974, but in another form.

The "colicensee proposal" referred to two licensees, licensed in tandem, sharing the same basic cable revenue. One licensee would operate the distribution system, the other would provide a new professionally produced channel. In the "universal pay television" arrangement, cable licensees would be required by regulation to carry the service provided by a separate programming licensee, which would charge the former for the service. The level of payment would be regulated by the CRTC. The two schemes, for all practical purpose, worked the same way.

Pay-TV American-style duplicated the American-dependent economic structure of conventional commercial television in Canada. Both also had extraordinarily wasteful financial overheads. Both also required large transfers of funds to the United States. The use of the general cable subscription mechanism already in existence, on the other hand, had no incremental financial overheads, was independent of additional American programming, required little or no net transfers of funds to the United States and elsewhere, required relatively little expenditure on marketing and drew on a source of revenue directly available up to a high ceiling and on a regular utility basis. A substantial part of the proposed allocation, moreover, could be recovered from existing excessive cable profits, as the proponents of the universal channel were quick to point out. Unlike pay-TV U.S.-style, further, the cost of the universal channel per household would be negligible because of its universality, with everybody contributing. Equally important, universal distribution bypassed extra hardware expenses required by U.S.-style pay-TV to protect the signal from nonsubscribers. The Canadian programming, further, would be available through cable subscription to the better part of the population.

Universal pay television, in other words, was everything the rhetoric of the CRTC and the minister of communications asked for. Its Canadian programming capacity was also what the CRTC was directed by statute to implement. All this was basic and obvious, just as it had been in 1974. You could write down the essential argument for universal pay television on the back of a CRTC-hearing agenda.

But why did the CRTC not implement the obvious? Well, unlike U.S.-style pay-TV, universal pay television would provide no extra profit for cable holding companies. Nor would it introduce into households the descrambler box through which the companies hoped to market other services and make extra, unregulated profit. Universal pay television would also eliminate any excessive profit the companies were already collecting on the basic rate. In short, it did not appeal to the cable industry.

David Balcon, a communications consultant and commentator living in Edmonton, was the first to put the idea of universal pay television together and establish the model, while working for the CRTC, as it happened. Balcon was in charge of socioeconomic research at the commission. By the summer of 1976 he had prepared a background paper pointing out the weaknesses of the cable lobby's proposed pay-TV scheme and putting forward a well articulated universal pay-television model instead. Cable subscribers, under the proposal, would have a high-calibre entertainment-movie channel, with room even for A-rated features, plus children's and art films, plus concerts, plus hockey games, all for $1.65 per subscriber per month. The channel, on the basic service, carried to the cable companies by satellite, would also boost cable's importance and popularity (in a way, one can add, that a high-cost, fractional-audience subscription scheme could never do). An additional 35 cents per subscriber was envisioned for a television production assistance fund, to underwrite the production of Canadian "drama" series for broadcast by existing television stations.

At this point the Council of Canadian Filmmakers (CCFM) got involved. The CCFM, through its member-organizations, represented fourteen thousand people working in English-Canadian film and television production. The major work was done by Kirwan Cox and Sandra Gathercole. The CCFM undertook a three-pronged campaign to help the program production industry come to grips with the pay-TV issue. It published and distributed widely a pay-TV booklet. It organized a daylong seminar in Toronto, attended by two hundred people, on the impact of pay-TV on the program production industry, the transcript of which was submitted to the CRTC for its second pay-TV hearing in 1977. Finally it undertook a comprehensive research study which resulted in a 409-page report, *Pay-TV in Canada,* also submitted to the CRTC for the hearing. The report, written by Cox and Gathercole, was an exhaustive economic and technical investigation of pay television, including a detailed analysis of the U.S. pay-TV experience (most of the existing research and experience was American). The report also analyzed fifteen different proposals for pay-TV in Canada.

Pay-TV in Canada was one of the finest pieces of broadcasting-policy work ever done in the country, going back to the Aird Commission Report, and it came to the inevitable conclusion that the U.S. pay-TV model was inappropriate for Canada and that the "universal pay" model should be adopted.

"Having considered the options available," it explained, "we have concluded that this universal model would provide the optimum system for the introduction of pay-TV at this time. It would generate the largest possible revenue for Canadian production at the lowest cost to the subscriber. It would utilize the Canadian market by organizing it on the basis of maximum efficiency for a small market rather than on the American basis which is only appropriate for a large market."

The study also examined in some detail two existing universal pay television models, Gill Cable in San Jose, California, and Videotron on the south shore of Montreal. Gill Cable had piggybacked a movie channel onto its basic service with a $2-per-month rate increase. "The development of pay-cable [on a nonuniversal basis] had been all wrong," Gill Cable's consultant described the situation. "The 'black box' [descrambling hardware] syndrome dictates the economics, and the lease channel/black box approach is poor economics."

In the mid-1970s Videotron offered, as well as the usual off-air channels, ten data information channels and eight demand programming channels (where viewers could call in and select programs from a central library). Videotron spent at the time over one-third of its annual budget on these local "universal pay television" channels and, like Gill Cable, used the extra service to attract more subscribers.

So, by the time of the 1977 hearing on pay-TV, the CRTC had before it, in addition to the internal working paper, a comprehensive study examining all the options and putting forward the only one that stood up. This proposal was backed by a coalition including the CCFM, the Canadian Conference of the Arts, the Canadian Broadcasting League, Selkirk Holdings, the Canadian Labour Congress, the Committee for an Independent Canada, the Canadian Film and Television Association, NABET (the television technician's union), the Union des artistes (French-language performers) and the Syndicat général du cinéma et de la télévision. In the commission's report on the hearing the following spring, the CRTC called the CCFM brief "compelling." The commission also cited nationalist declarations from the Aird Commission. It quoted R. B. Bennett at length. It cited the Massey Commission. It cited the Fowler Commission. It cited the late Fowler Committee. It cited at length previous noble declarations the CRTC had made. It quoted vastly from the pious declaration on pay-TV made by Jeanne Sauvé in 1976 and from the follow-up pious declaration on pay-TV made by CRTC chairman Harry Boyle. It was another typical CRTC document.

But not to be forgotten was the Third Law of Regulatory Agencies: The more elaborate the ritualistic language used by the agency, the more dubious are its actions. When it came to taking up the universal pay television proposal — to actually doing something with it — the CRTC ducked. Its report said that the "universal" element in the proposal "requires a mandatory tax on all cable subscribers that is not sanctioned by legislation." Mind you, this was not true. The universal channel was a proposal to amplify the basic cable package, well within the commission's authority. It was CRTC business as usual.

Even if the commission did not have the legal authority, common sense should have led it to turn the matter of pay-TV over to Parliament with a forceful recommendation for universal pay television, while refusing to act on pay-TV otherwise. A regulatory agency should not resort to bad policy choices simply because it does not have the necessary jurisdiction itself.

The next CRTC hearing on pay-TV, the third, conducted by the Therrien Committee, took place in 1980. A Joint Action Committee (JAC), with a slightly different membership, picked up the work of the 1977 CCFM coalition. The JAC put forward a comprehensive universal proposal and once more patiently went over the structural folly of U.S.-style pay-TV in Canada. The JAC proposal called for a $2.50 allocation per subscriber towards Canadian program production in a basic Canadian cable package including new services like CBC-2, a National Film Board–House of Commons channel and a northern native service. This was to ensure a floor of Canadian content in the oncoming wave of satellite service to northern and remote communities, which the Therrien Committee was considering at the same time.

The JAC's proposal also included a $1.50 allocation per subscriber towards English and French "premium" entertainment channels. These fitted the narrowly defined "pay-television" category. These premium channels would be optional to each cable system, and therefore "discretionary" in that sense, but once taken they would go to all of that system's subscribers, again bypassing the expensive hardware and other overhead costs of U.S.-style pay-TV. The coalition was able to cite existing precedents for universal fees of this sort. The CRTC, for example, had authorized a 50-cents-per-subscriber fee increase for Quebec cable systems choosing to carry a packaged channel of programming from France, to cover transmission costs. The JAC was also able to cite earlier statements by Canadian Cable Television Association officials (the cable lobby) actually proposing universal allocations for new Canadian programming on cable.

The Therrien Committee ignored the option, and any variation of it, in its recommendations.

For the next CRTC hearing on pay-TV, the fourth, which was the 1981 licence hearing, the CRTC's call for applications deliberately tried to exclude "universal" applications from being heard. A new incarnation of the support for the concept, under the name of TeleCanada, fought back. TeleCanada was originally organized by Jack Gray, a long-time president of ACTRA. Gray was succeeded by Toronto consultant and policy analyst Paul Audley, a former executive director of the Association of Canadian Publishers. Audley had not dealt with the CRTC before. As a result he gave its pious protestations some credence. He was inspired, if not thunderstruck, by the fact that the universal cable proposal made so much common sense. Why had it not been implemented before? TeleCanada, by bending ears, convinced the commission that it should hear the application.

The TeleCanada application promised a minimum of 75 per cent Canadian content exclusive of repeats, and the expenditure of 80 per cent of gross annual revenue on the purchase and commissioning of independently produced Canadian films and programs. All of the $2.50-per-month subscriber allocation would go into this broadcasting organization, rather than being creamed and milked off the top, up to 45 or 50 per cent, by the cable companies, for special hardware and for profits, which would be the case for U.S.-style pay-TV. TeleCanada's not-for-profit character would, further, maximize the allocation to programming, while its assured revenue would free it to concentrate on its Canadian programming mandate.

The 80 per cent expenditure on Canadian production compared to the minor expenditure from total revenue of U.S.-style pay-TV, which in turn could be spent on Americanized phony-Canadian programming for sale in the U.S. Touching on the crucial point in an introductory section, the application pointed out that TeleCanada would be "the first Canadian private sector programming service to have no inherent commercial incentive to reduce its investment in Canadian programming or to focus on denationalized programs or to expand its use of less expensive and less risky foreign programs."

TeleCanada's Canadian programming expenditures were projected at $150 million a year, all from $2.50 per month per subscriber. The $2.50 per month also covered non-Canadian program acquisitions and satellite and administrative costs. This programming expenditure would finally put the Canadian program production industry on a firm footing. The single largest category of projected Canadian programming expenditure was $50 million annually for feature films. The second-largest category, $44 million annually, was for specials. The TeleCanada channel would be primarily an entertainment channel, but with children's programming and a variety of other elements.

The TeleCanada group was impressive, too, in terms of playing CRTC games. Its sponsors included Walter Gordon, Pierre Berton, Allan King (the film and television director), Jack Biddell (partner in Clarkson Gordon), Kealy Wilkinson (broadcasting policy consultant, in the mid-1970s executive director of the Canadian Broadcasting League), Micheline Lanctot, Lloyd Shaw, Phyllis Lambert, Claude Jutra, Abe Rotstein, Christina McCall-Newman and former communications minister David MacDonald.

The TeleCanada application envisioned a voting membership of twenty-five broadly representative people, including the founding members and others invited "on the basis of their capacity to contribute to the business and affairs of the corporation." These members would elect a board of fifteen directors to manage and direct the corporation, of which five members, for at least the first two years, would represent the CBC, National Film Board, Canadian Association of Broadcasters, Canadian Cable Television Association and the Canadian Film Development Corporation.

TeleCanada differed from previous universal proposals in one respect. It did not oppose the licensing of nonuniversal applicants, channels inevitably centred on American movies, and where subscribers would have the option of taking the service or not. This gave the cable lobby what it wanted while at the same time trading it off for a more Canadian service to balance the American pay-TV. It did the same for cable subscribers. The universal, nonoptional contribution to the Canadian channel would be twinned with the availability, if the subscriber so desired, of an American-movie pay-TV channel or channels. Finally, that part of the $2.50 per month not available from excessive cable profit margins would hardly be noticed when compared to the $15 or more charged by the nonuniversal services. David MacDonald went out of his way at the 1981 licence hearing to point out to the CRTC the value of licensing the universal service at the same time, and not delaying it.

The brief encounter between MacDonald and CRTC chairman John Meisel, at the hearing, had so many ironies that once you understood them, and how they came about, you understood everything. MacDonald, with Joe Clark, had appointed Meisel because of Meisel's reputation as a nationalist. Yet Meisel was about to perpetrate a screwball Americanizing pay-TV scheme, which MacDonald opposed, while at the same time resisting the only scheme presented at the hearing that made Canadian sense, the TeleCanada proposal, which MacDonald, no longer minister, was pleading for. But how had this hearing come about? It had come about as a result of the Therrien Committee formed by the CRTC at the instigation of then Minister of Communications David MacDonald to hurry pay-TV along. MacDonald, at the pay-TV licence hearing, found himself opposing a development he had helped to bring about.

The problem, MacDonald realized, with a wry sense of humour, was that it was not until he stopped being minister and was on the outside looking in that he really understood what was happening. While he was minister, he was a prisoner of the process, trussed up by the paper he was being fed. It was, it seemed, something right out of the BBC comedy series *Yes, Minister,* where the Cabinet minister learned the hard and funny way just how much of a prisoner he was. MacDonald, however, now did understand what the process was doing to Canadian television. But unfortunately for him, Meisel did not. He was still part of the process. Nothing Mac-Donald could say in a public-hearing exercise would alter that. Meisel's pay-TV fiasco then continued to unfold, in all its absurd-ity, as the former minister of communications who appointed him watched in astonishment.

While awarding the U.S.-pay-TV-in-Canada licences, the CRTC refused to decide on the universal concept. Although claiming at length to find favour with the universal concept, the commission raised the question of how the licensee could be held accountable for the nature and quality of its programming. This was a wonder-ful question from an agency that had failed to hold its private television licensees accountable but kept renewing their licences anyway. The secure revenue source in the TeleCanada case was exactly what would allow the licensee to do what it promised to do.

But with the secure revenue source, what would prevent the licensee from ignoring the public in its programming? This raised the terrible spectre of a small group indulging their own idiosyn-cratic tastes using the public's millions. On the other hand, why would the licensee want to ignore the public when its purpose was to attract the public? Moreover, because of the nonprofit character of a TeleCanada licensee and its minimal assets (it would not pro-duce its programming itself and hence not maintain a production plant), the licensee could easily be changed when the licence expired. There would be so much attention focussed on the new network, for which every cable subscriber was paying, that pressure from the public alone would keep it accountable.

Equally obvious, the commission could have issued the licence conditional on a new governing structure being approved. This would have put "universal pay television" in place, where imple-mentation of the idea could not be sabotaged.

Instead, the CRTC announced there would be another hearing on universal pay television. The TeleCanada principals were shaking their heads at the missed timing. Word leaked out from supporters of the universal proposal within the commission that the delay was not critical. The principle had been accepted, they said, as part of the original pay-TV decision. It was no longer a matter of if, only a matter of when.

The cable lobby now, for the occasion of this next hearing in late 1982, hammered away at the mandatory aspect of universal pay television. This was the original bogey. The CRTC would be taking subscribers' money and forcing the channel down people's throats whether they wanted it or not, the lobby said in so many words. Phil Lind, vice-president of Rogers Cablesystems, the heavyweight after all those takeovers, told the commission that such a terrible deed would bring the government down. Since most members of the commission were good Liberal appointees, this was a nice touch. Rogers, to the CRTC, represented power. Mind you, this power was created for Rogers by the CRTC in the first place.

A front-page story by Jack Miller, Toronto *Star* broadcasting writer, contributed to the alarm. Miller lumped in "universal pay television" with the possibility that cable subscribers would have to pay extra for the U.S. channels they were now getting. "Forced," "powder keg" and "explosive potential" were phrases well featured in the story.

The cable lobby at the same time pressed for a new tier of American specialty channels, to be packaged at an extra price, from which revenue it would allocate 15 per cent to add Canadian elements. The lobby again raised the spectre of satellite receivers threatening cable, in a few years for homes as well as apartments. Canadians would disconnect from cable *en masse*, went the scenario. Canadian broadcasters might lose access to Canadian viewers ("an alarming possibility," said Rogers Cablesystems) because of this Trojan Horse. The vital Canadian role of cable, as the cable lobby described it — program substitution revenue, expanding the audience of UHF-stations — would be bled away. The clock was ticking. The only way to stop this crumbling of the national system was to license cable for U.S. specialty channels.

It was that old Canadianization-through-Americanization scheme again.

The righteous denunciations of "mandatory imposition" were not exactly based on a disinterested analysis of how television was paid for. To start with, cable, of which the proposed service would be a part, was discretionary, not mandatory. Cable, indeed, *was the only discretionary part* of the broadcasting structure. All private commercial television and CBC commercial revenue was financed by mandatory payment; people contributed involuntarily by participating in society as consumers whether they wanted the commercial television or not.

Within cable the charge for the cost of upgrading systems to handle augmented capacity was universal, imposed by cable companies on every subscriber whether the individual subscriber had a converter or not, or wanted to pay for the extra channels or not, or wanted to use the extra capacity for pay-TV or not. The community channel was universal. Payment for other cable-originated channels was universal. Payment for Rogers's "legislature service," that generously provided video facilities to legislators, was taken unilaterally from the backs of subscribers. Later, Ted Rogers would propose a universal charge for basic subscribers (he called it a rate increase) to subsidize specialty satellite channels that he was after, including privately operated channels presumably with commercials.

Excessive cable rates were universal, and subscribers did not get any programming at all for these payments, just a cable lobby and more rate-increase applications. The proposed universal cable channel would have converted that wasted payment into programming. Cable companies drew on the mandatory payment they took from subscribers to lobby against this conversion, with dark cries about mandatory payments.

The APBBC also attended this 1982 hearing. The association, of course, happened to have a real solution to the CRTC's imaginary accountability problem. *Let the subscribers govern the network.* They are paying for it, why not let them govern it? A subscriber-governed network was something that should have been implemented in its own right. As the APBBC had realized from the beginning, the cable subscription mechanism, by rare historical luck, allowed for the creation of a media sector owned and controlled by the citizenry themselves, through membership organizations.

For a timid CRTC, stricken by the cable lobby and fading fast, the idea of a "cable subscriber network," as the APBBC called it, was a godsend, too. A network that viewers themselves would own and govern was a powerful populist selling point, particularly against cable holding companies.

The commission, naturally, ducked again. Its response to the hearing was delayed for almost a year while the prospect of action faded. In the document finally issued, October 1983, it avoided altogether discussion of a subscriber-governed network. In a resumé of the proposals, it avoided even mentioning a subscriber-governed structure. At the same time, it claimed that problems of accountability still had not been solved!

The CRTC therewith put off indefinitely the introduction of a cable-subscriber network. In the same document it announced the establishment of a consultative committee to look into the possibilities. This committee, the announcement said, would be comprised of "representatives from the cable and television broadcasting industries and other interested parties" appointed by the commission. Mind you, the cable and broadcasting industries had opposed the idea of the new network. Moreover, there was nothing that a consultative committee could contribute that the regulatory agency could not, and should not, find out independently for itself. Who was making the decisions anyway?

On the very same day as the CRTC's announcement on the universal channel, someone from the industry, André Bureau, was appointed chairman. Meisel had already announced his intention to leave, for personal reasons.

Bureau was president of Canadian Satellite Communications (Cancom), a consortium principally owned by BCTV, Allarco, Telemedia and Selkirk Holdings, distributing signals by satellite to northern and rural areas. Before that he was president of Télémédia (the de Gaspé Beaubien broadcasting holdings), and had acted in that capacity in one of the unsuccessful pay-TV applications. The partner in the application, Harold Greenberg (Astral Bellevue), was about to take over First Choice with Edward and Peter Bronfman. When Pierre Camu, the president of the Canadian Association of Broadcasters, was appointed CRTC chairman in 1978, there were denunciations approaching furor about putting the fox in charge of the chicken coop. Now, nobody bothered.

Meanwhile, high-cost U.S.-pay-TV-in-Canada flogged its way forward.

Running just ahead of and parallel to the pay-TV decision was the decision to license a satellite service for northern Canada and isolated rural areas elsewhere, a licence granted to Cancom from whence came Bureau. The purpose of this licence was to head off the inroads of U.S. satellite signals being picked up by pirate operators of satellite dishes and to protect the North for Canadian broadcasting, while offering wider choice. It was also supposed to "respond to the particular needs" of these remote areas, to cite the CRTC's pious language in the call for applications.

Only one Canadian channel in each language, the CBC English and French, was previously available in these outlying areas. But the idea of an expanded eleven-channel Canadian package, including a native service, as the base of extended satellite and cable carriage — the idea developed by the Joint Action Committee for the whole country — got left in limbo. The CRTC's licensee Cancom provided, instead, BCTV (Vancouver), CHCH (Hamilton), CITV (Edmonton) and a French schedule from Montreal. The core schedules of the English-language stations just happened to be

American. The late afternoons were American reruns. No expenditure for native programming in the North was required, only the provision of ten hours a month of satellite distribution time for native programming financed from other sources. Viewers in Frobisher Bay and Ellesmere Island, however, could get the local Hamilton and Edmonton news.

When additional satellite capacity became available, Cancom, as envisioned, applied to carry one PBS and three American commercial stations, just like southern cable systems. If U.S. channels were carried by microwave from distant head-ends to places like Edmonton and Prince George, why should they not also be carried by satellite to far-northern points? Viewers in Frobisher Bay and Ellesmere Island could now get the local Detroit news as well as the local Hamilton and Edmonton news. There was nothing quite like these schemes for protecting the goals of Canadian broadcasting.

Later, in the spring of 1984, with ex-Cancom president Bureau installed as CRTC chairman, the commission announced it would allow cable licensees to carry five U.S. specialty channels which they could select from a long list. The list included the Cable News Network — the commission leaving the all-news territory, for the time being, to American control. At the same time, the commission licensed a specialty music video service (CHUM Ltd.) and a sports network (Labatt's) with partial Canadian content beginning at 10 per cent and 18 per cent respectively. This was offset only by greater Canadian content on the sports network during prime-time evening hours (6–12 P.M., 34 per cent) and peak evening hours (7:30–10:30 P.M., 47 per cent) taken alone. If a cable system distributed the maximum allowable of these newly approved specialty channels, the average Canadian content added to its carriage would be a glorious 4 per cent for the first two years and, if feasible, but not necessarily feasible, a maximum 9 per cent down the line.

The Canadian Cable Television Association was now in full, well-oiled effulgence, despite the hard times in the economy. It had fifteen employees and an annual budget of $1.3 million. Lobbying activities and expenditures by individual companies would be on top of that. One of the regular CCTA events was the annual "MP's reception," for members of Parliament and other Ottawa officials. One former MP remembers it as being second in quality only to a reception thrown by some Arab oil states.

Still later, in the summer of 1984, First Choice and Superchannel gave up the ghost of competition, splitting the territory. First Choice took eastern Canada, Superchannel took the West. They had run up $40 million in debt in their first eighteen months of operation.

Of the original pay-TV operations, licensed in 1982 and 1983, First Choice had now been forced to change hands and had lost half the country, Superchannel had sacrificed its Ontario licence, C Channel had gone bankrupt, Star Channel had gone bankrupt, Aim Satellite in B.C. failed to get off the ground, Premier Choix and TVEC had merged in desperation, and the small multilingual operator in Vancouver, World View Television, had been thrown into receivership.

4

Rogers Takes Over
IN MORE WAYS THAN ONE

IN BRITISH COLUMBIA Rogers Cablesystems, as the Rogers corporate vehicle was now called, had been harnessing the newly acquired British Columbia licences to the Americanizing corporate strategy.

Although at the Premier takeover hearing in 1980 George Fierheller, president of Premier, on behalf of the takeover parties, had solemnly expressed the belief "that it is very important that [the] local management remain in place," one of the first things Rogers Cablesystems did, with the licences in hand, was get rid of the veteran general manager of the Victoria system. He had run the system well, but he did not have the slick public-relations skills or the executive style that fitted Rogers's corporate requirements. Rogers replaced him with the young general manager of the small Chatham, Ontario, system who did fit to type. Rogers then dissolved Victoria Cablevision Ltd., Vancouver Cablevision Ltd. and Fraser Cablevision Ltd. (Coquitlam), making them internal divisions of Premier.

The high rates of return from the Victoria and Vancouver systems were applied towards the inflated Premier takeover cost. Expansion in the United States was also eating up returns. The total assets of Rogers Cablesystems in 1979, before the takeover of Premier, were $120 million. At the end of 1983 and U.S. expansion, they were $914 million. Long-term debt rose accordingly from $15 million in 1979 to $637 million in 1983. The cash flow from Rogers's Canadian cable systems was expected to help support and amortize this indebted financial superstructure.

The dissolution of the separate corporate entities meant that audited financial statements for the individual licensee operations were no longer prepared for the registrar of companies. Rogers

Cablesystems thereupon began filing with the CRTC audited statements only for its incorporated holding companies. This made proper rate regulation impossible. The commission's machinery swallowed the omission without a peep. The commission itself, at rate-increase hearings in 1982 and 1984, rejected all motions that audited statements for the individual licence operations be made available, despite its own 1975 policy calling for them.

Behind this screen, Rogers Cablesystems manipulated the individual licensees' financial statements, which were unaudited, in attempts to get higher rates. To use Victoria as an example, the rate base on which rate of return is measured was artificially tripled in a way that was caught only because Capital Cable Co-operative, in 1982, awaiting the resolution of a final court action on the takeover, was still on the scene. At that, the manoeuvre was detected only by accident and because of previous knowledge of the system. The CRTC had noticed nothing and had no idea what was happening.

Simultaneously, expenses on which rates were calculated, that is, the expenses as listed in the financial statements, skyrocketed. Administration expenses for the Victoria system increased 155 per cent (to two and a half times the previous level) over the short three-year period 1980–83 since the takeover, with only a minimal increase in the number of subscribers. The Vancouver and Coquitlam figures were much the same. Motions were made at the two rate-increase hearings in 1982 and 1984 for regulatory audits to determine which expenditures were legitimate for rate regulation purposes and which had been applied to other activities and non-competitive intracorporate transactions. These motions were all rejected by the CRTC.

The figures in these financial statements were prepared in Toronto. The function of the local managers was to serve as front men for the rate-increase applications based on the figures. At the same time, in 1982, a batch of community channel employees was laid off by order of the parent company. The "legislature service," promised at the takeover hearing, was suspended. The excuse given was that with the federal government's six-and-five restraint formula, which applied to rate increases, the rate of return was inadequate, but behind the accounting manoeuvres, the reality was quite different. Corrected financial statements prepared for the City of Victoria by C. L. Mitchell of the Faculty of Commerce and Business Administration, University of British Columbia, rectifying the rate base and scaling back "administration" expenses to appropriate levels, showed that the rate of return was, and had been, excessive as before. These corrected statements were hotly contested by Rogers Cablesystems. The eloquent declarations at the takeover hearing about preserving "the tradition of decentralized management," meanwhile, were a dead letter.

The cutting of community programming staff and of the legislature service was a method of unilaterally increasing the profit margin without the impediment of a public hearing and public scrutiny — a rate-of-return increase off the bottom, bypassing rate regulation at the top. Altogether Rogers, through its various systems, laid off 175 employees.

An earlier, inadvertent disclosure by Premier president George Fierheller revealed that the administrative costs of an unrelated takeover attempt by the parent company were charged back to licence operations, indicating that attempts were being made, and would be made in the future, to get subscribers, through increased rates, to pay the costs of all similar unrelated corporate activities and politicking.

An army of officers, including the company's "outside" lawyer from Ottawa, were thrown into the rate-increase hearings to protect these manoeuvres. In this fashion the cable holding company could denounce a new Canadian channel at $2.50 per month per subscriber, where over time it might manoeuvre past the agency unjustified rate increases of a similar scale to feed its growing empire. As it happened, the CRTC in mid-1984 awarded $1.50-per-month rate increases to Rogers for its Victoria and Vancouver systems, and a $1.37-per-month increase for Coquitlam, ignoring the accounting manipulations and the multiplying "administration" expenses. Applications for a subsequent set of increases, already indicated in the 1984 proceeding, followed for the next year — another $1.50 per month for Victoria and Vancouver, half that for Coquitlam whose rate was higher.

Meanwhile, Rogers, with 25 per cent of all cable subscribers in Canada and 30 per cent of all English-speaking cable households, including key systems in Vancouver and Toronto, now had a hand on the gate to distribution. Nobody planning a "discretionary" satellite network could proceed until they had Ted Rogers's blessing or could convince the CRTC to make him accept their terms.

We now have the end picture of the closed takeover — a hierarchical and homogenizing corporate structure; centralized financial decision-making; centralization of services; escalating "administration" expenses; unabashed attempts at manipulation of a feeble regulatory agency ("protecting our investment" is the euphemism);

heavy expenditure on regulatory games; "gatekeeper" leverage; the subordination of the local licence operation and of local cultural possibilities, and the use of the resources from the panoply of licences to build up an Americanizing marketing power and to block Canadian initiatives like a cable subscriber network (universal pay television). It all hinged on control of local licences. There might be everlasting hearings and mountains of submissions on universal pay television or on any other cable policy subject you may want to name, but the root power in the game, as Capital Cable Co-operative knew, was the financial and administrative control of the local cable licences.

5

The Joke Is on Us All

WATCHING EVENTS IN the early 1980s was like watching the Toronto remake of an old British Columbia movie. A growing number of people in deepest Toronto began to realize, from the accumulating detail of the pay-TV and universal pay television episodes, the true nature of the CRTC, the agency fathered by Juneau and Boyle. The old, received idea of the CRTC — that it existed to implement Section 3 of the Broadcasting Act because Section 15 of the Broadcasting Act said so — did not survive the experience.

In 1982 film director Allan King told the commission that "year after year, now becoming decade after decade, we see this gross disparity between good intentions and terrible performance ... year after year your decisions have magnified the problem."

In 1983 Robert Fulford in *Saturday Night* began writing less than flattering things about Pierre Juneau, quite a difference from the hosannas the magazine had indulged in, in the early 1970s.

During this period the regulatory and policy-making players ended up doing just about what you would expect them to end up doing:

Former CRTC vice-chairman Charles Dalfen was practising law in associaton with the firm of Johnston and Buchan, well-known communications lawyers. Dalfen acted for the ill-fated pay-TV syndicate Lively Arts Market Builders (C Channel) and was on the board of the Cable Telecommunications Research Institute, the cable industry's research organization.

Chris Johnston, of Johnston and Buchan, had been CRTC general counsel in the days when Capital Cable's first battles for competitive hearings took place. Johnston was retained by the Ottawa-financed Law Reform Commission of Canada to do a study of CRTC procedure, later issued in book form. (The reform commission in Ottawa saw nothing amiss in a former staff member doing the study on his old Ottawa agency.) Except for a passing legal note, Johnston excluded from the study the commission's refusal to hold competitive hearings when licences expire or change hands, as fundamental and conspicuous an issue of procedure as anybody could ask for. This was quite a feat. Among the clients of Johnston and Buchan were the Canadian Cable Television Association and Rogers Cablesystems.

Roy Faibish, the CRTC commissioner who had been party to the commission's approval of the Rogers takeover of Premier and then joined the Rogers organization, spent a couple of years as "vice-president — Europe" in the Rogers Cablesystems corporate hierarchy. "At least," said one wit when Faibish joined Rogers, "he had the good sense to get out of the country." Faibish suffered his exile in London, England.

Jacques Hébert, CRTC part-time commissioner from 1970 to 1980, and friend of Pierre Trudeau, was cochairman of the Applebaum-Hébert Federal Cultural Policy Review Committee. Hébert was the one who always got indignant about the new private stations' broken promises, while faithfully sitting on the commission which had licensed them and which regularly renewed the licences. The Applebaum-Hébert Committee's report, however, came down hardest on the CBC, which was at least approximately doing its job. It also recommended the licensing of more private stations, under strict regulatory control, of course, to ensure that these new services contain almost exclusively Canadian programs (just like City-TV's original licensing, no doubt). The committee did not offer a critical examination of pay-TV. It did not recommend a CBC-2, which conflicted with the plans of the Lively Arts Market Builders pay-TV syndicate (C Channel). The other cochairman of the committee was Louis Applebaum. When the committee's work was done, Applebaum returned to the board of the Lively Arts Market Builders pay-TV syndicate from whence he had come.

Francis Fox was still minister of communications, making great progress in vacuous speeches with titles like "Broadcasting — A New Era," "Broadcasting and Cultural Sovereignty" and "Pointing the Way To A Strategy."

Harry Boyle, in semiretirement, wrote profound articles and made profound speeches. In one he defended the idea of universal pay television, insisting with comparative illustrations that its financing through cable was not a tax. (Flashback: Boyle exited as chairman in September 1977 without sticking his neck out in defence of universal pay television; six months later, the new regime's report on pay-TV set aside the proposal because, they announced, it involved a tax.)

In another article Boyle lashed out at editorial access to the media being available "simply to those who have the resources to pay for it." He was particularly indignant about "government advocacy." "When [government] try to end run the process of the media as well, they are taking undue liberties with the democratic form." (Flashback: "The commission is not convinced that it is desirable at this time to hold a public hearing into the general question of the appropriate use of advertising time," said the CRTC, under Juneau and Boyle, with regard to advocacy advertising. The right time never came.)

"The premise of using [pay-TV] revenue derived from the showing of primarily American movies to cultivate a new cornucopia of Canadian films and television is a dubious one," Boyle wrote. (Flashback: "An important and serious first step towards the repossession of a Canadian broadcasting system," said CRTC chairman Boyle of pay-TV possibilities in 1976.)

"We can talk all we want about our current situation but the real test of our commitment is what we do," said former CRTC chairman Harry Boyle.

Jean Fortier, like Dalfen a former CRTC vice-chairman, and the one who so elaborately skated around "advocacy advertising," turned up as vice-president of the Premiere pay-TV syndicate. Fortier became particularly indignant at criticism of his syndicate by the Institut canadien d'éducation des adultes, published in Le Devoir. If the people do not want pay-TV, he argued back with fervour, "it will be dramatic only for us because we will go bankrupt." When Premiere failed to get the national licence, Fortier became president of TVEC, which was subsequently awarded the French-language regional licence. The outfit did plunge quickly towards bankruptcy, upon which disaster Fortier ran dramatically to the Quebec government for rescue money.

The Liberal backbencher and ex-CRTC researcher who talked out James McGrath's bill to abolish television advertising directed to children, namely Monique Bégin, became minister of health and welfare, including the health and welfare of children.

The other Liberal backbencher who participated in that exercise, Mark Raines, was appointed a part-time member of the CRTC.

John Meisel returned to Queen's University, which offered him the distinguished Peacock Chair. His work there, read the CRTC resignation announcement, was "expected to focus on cultural policy, the politics of regulation, and the political implications of the information society."

"I shall inevitably draw heavily on the experience and knowledge I have gained at the CRTC," said Meisel.

"John Meisel should be pleased that he has contributed to the possible realization of that vision [the reinforcement of a Canadian presence in broadcasting]," wrote *Maclean's,* true to its journalistic fantasyland to the end.

Pierre Juneau was president of the CBC. Many people quickly noticed that, in defending the CBC against calls for radical change made by the Applebaum-Hébert Committee, Juneau was taking the position of Laurent Picard, CBC president in 1974, who was defending the CBC against calls for radical change by CRTC chairman Pierre Juneau, whose arguments nicely resembled the ones Juneau was now opposing. Shortly after Juneau's appointment to the presidency, he was honoured at a dinner in the York Club, thrown by his old friend John Bassett, chairman of Baton Broadcasting (CFTO et al.).

Jeanne Sauvé, the lady who gave us the Saskatchewan cable fiasco and, in its origins, the pay-TV debacle, and who denied Capital Cable Co-operative's appeal for the right to apply, was named Governor General.

As for Ottawa, it was where it always was, 248 miles northeast of Toronto and 126 miles west of Montreal.

Index